Taming Liquid Hydrogen:
The
Centaur
Upper Stage Rocket
1958–2002

Virginia P. Dawson and Mark D. Bowles

The NASA History Series

National Aeronautics and Space Administration
Office of External Relations
Washington, DC 20546
2004

NASA SP-2004-4230

Library of Congress Cataloging-in-Publication Data

Dawson, Virginia P. (Virginia Parker)
    Taming liquid hydrogen : the Centaur upper stage rocket, 1958-2002 / Virginia P. Dawson,
    Mark D. Bowles. p. cm. (NASA-SP-2004-4230)
    Includes bibliographical references and index.
    1. Centaur rocket—History. 2. Hydrogen as fuel—Research—United States—History. 3.
    Liquid propellant rockets—Research—United States—History. I. Bowles, Mark D. II. Title.

TL783.4.D39 2004
621.43'56—dc21

                                                                    2004042092

Table of Contents

Introduction

During its maiden voyage in May 1962, a Centaur upper stage rocket, mated to an Atlas booster, exploded 54 seconds after launch, engulfing the rocket in a huge fireball. Investigation revealed that Centaur's light, stainless-steel tank had split open, spilling its liquid-hydrogen fuel down its sides, where the flame of the rocket exhaust immediately ignited it. Coming less than a year after President Kennedy had made landing human beings on the Moon a national priority, the loss of Centaur was regarded as a serious setback for the National Aeronautics and Space Administration (NASA). During the failure investigation, Homer Newell, Director of Space Sciences, ruefully declared: "Taming liquid hydrogen to the point where expensive operational space missions can be committed to it has turned out to be more difficult than anyone supposed at the outset."[1]

After this failure, Centaur critics, led by Wernher von Braun, mounted a campaign to cancel the program. In addition to the unknowns associated with liquid hydrogen, he objected to the unusual design of Centaur. Like the Atlas rocket, Centaur depended on pressure to keep its paper-thin, stainless-steel shell from collapsing. It was literally inflated with its propellants like a football or balloon and needed no internal structure to give it added strength and stability. The so-called "pressure-stabilized structure" of Centaur, coupled with the light weight of its high-energy cryogenic propellants, made Centaur lighter and more powerful than upper stages that used conventional fuel. But, the critics argued, it would never become the reliable rocket that the United States needed.

Others, especially military proponents of Centaur, believed that accepting the challenge of developing liquid-hydrogen technology was an important risk to take. Herbert York, Chief Scientist for the Advanced Research Projects Agency (ARPA), had urged NASA Administrator T. Keith Glennan to expedite the program in 1959 because Centaur was "the only vehicle that has the capability of meeting our payload requirements for high altitude orbits."[2] Six months after

---

[1] U.S. House of Representatives Subcommittee on Space Sciences of the Committee on Science and Astronautics, "The Centaur Program," 87th Congress, 15 and 18 May 1962 (hereafter cited as Centaur Program, 1962 Hearings), 11.

[2] Herbert York to T. Keith Glennan, 19 September 1959, Centaur files, NASA Historical Reference Collection.

Centaur's aborted first flight, President John F. Kennedy demanded to know what NASA hoped to achieve with Centaur. Administrator James Webb's response was unequivocal. With the Apollo program already committed to using liquid hydrogen in the upper stages of the giant Saturn vehicle, Centaur would prove the fuel's feasibility.[3] The Agency was also counting on Atlas-Centaur to launch Surveyor, a robotic spacecraft with a mission to determine whether the Moon's surface was hard enough to land future spacecraft with human beings aboard.

Despite criticism and early technical failures, the taming of liquid hydrogen proved to be one of NASA's most significant technical accomplishments. Centaur not only succeeded in demonstrating the feasibility of liquid hydrogen as a rocket fuel, but it also went on to a brilliant career as an upper stage for a series of spectacular planetary missions in the 1970s. Ironically, this success did little to ensure the future of the Centaur rocket. Once the Shuttle became operational in the early 1980s, all expendable launch vehicles like Centaur were slated for termination. Centaur advocates fought to keep the program alive. They won funding for the redesign of Centaur as an upper stage for the Shuttle, spent nearly $1 billion integrate them, and then witnessed the cancellation of the program within weeks of the first scheduled flight of Shuttle/Centaur.

Miraculously, Centaur survived into the commercial era of the 1990s and is still flying as the upper stage for the Atlas. Although unthinkable at the height of the Cold War, the idea of privatizing the delivery of launch vehicle services gained currency in the early 1980s because it dovetailed with the free-enterprise, small-government ideology of the Reagan administration. Now NASA is just a customer—albeit a favored one—of a new service that rocket manufacturers offer to a variety of customers, including foreign governments. A competitor of the European Ariane rocket and the Boeing Delta, Centaur continues as the upper stage for a redesigned Lockheed Martin Atlas.

Centaur's importance in the history of rocketry has escaped most historians of the space program. One reason is that upper stage rockets compose the murky middle phases of space-flight that rarely make news headlines. They do not create the dramatic plumes of fire that the public sees on the launch pads at Kennedy Space Center. Nor do they carry space probes all the way to the distant planets. The booster or first stage of a multistage rocket uses the brute force of its large rocket engines to propel the stages and payload through the atmosphere. Once the booster has used up its propellants, it is jettisoned rather than burdening the remaining stages with the weight of a spent rocket. At this point, Centaur takes over. The role of the upper stage is brief—usually about 10 minutes of glory between the shutdown of the booster engines and the release of the spacecraft or satellite. Centaur's final task is simply to get out of the way, while the payload—the spacecraft—is just beginning its (often multiyear) journey.

---

[3] Transcript of Presidential Meeting in the Cabinet Room of the White House, 21 November 1962, 19, tape no. 63, John F. Kennedy Library President's Office files, NASA Historical Reference Collection.

Despite this neglect, Centaur's importance to the space program, and satellite communications in particular, is unquestionable. Centaur serves as the critical link between its booster stage (Atlas or Titan) and the mission's payload (satellite or spacecraft). The sole objective of the Centaur is to add the extra speed needed to guide the payload into a desired orbit and to orient it before separation. The more accurately Centaur does its job of positioning, the less need there is for the payload to use its own fuel to make up for any inaccuracies in trajectory. With this extra fuel, a spacecraft bound for a planet has more maneuvering capability both as it travels toward its destination and once it gets there. If the payload is a satellite, accurate positioning allows it to stay in orbit for a longer period of time. The ability of Centaur to restart its engines in space allows mission designers greater flexibility in accommodating the relative positions of the moving and rotating Earth and moving payload targets, whether planetary or lunar.

Though never identified with the dream of landing human beings on the Moon, or the product of a massive military crash program like the Atlas, Centaur has enjoyed an unusually long and sometimes controversial career as an upper stage rocket. As our title suggests, Centaur is especially notable because of its role in the development of liquid hydrogen as a rocket fuel. Hydrogen—a light and extremely powerful rocket propellant—has the lowest molecular weight of any known substance and burns with extreme intensity (5,500°F). In combination with an oxidizer such as liquid oxygen, liquid hydrogen yields the highest specific impulse, or efficiency in relation to the amount of propellant consumed, of any known rocket propellant.

Because liquid oxygen and liquid hydrogen are both cryogenic—gases that can be liquefied only at extremely low temperatures—they pose enormous technical challenges. Liquid hydrogen must be stored at minus 423°F and handled with extreme care. To keep it from evaporating or boiling off, rockets fuelled with liquid hydrogen must be carefully insulated from all sources of heat, such as rocket engine exhaust and air friction during flight through the atmosphere. Once the vehicle reaches space, it must be protected from the radiant heat of the Sun. When liquid hydrogen absorbs heat, it expands rapidly; thus, venting is necessary to prevent the tank from exploding. Metals exposed to the extreme cold of liquid hydrogen become brittle. Moreover, liquid hydrogen can leak through minute pores in welded seams. Solving all these problems required an enormous amount of technical expertise in rocket and aircraft fuels cultivated over a decade by researchers at the National Advisory Committee for Aeronautics (NACA) Lewis Flight Propulsion Laboratory in Cleveland.

Today, liquid hydrogen is the signature fuel of the American space program and is used by other countries in the business of launching satellites. In addition to the Atlas, Boeing's Delta III and Delta IV now have liquid-oxygen/liquid-hydrogen upper stages. This propellant combination is also burned in the main engine of the Space Shuttle. One of the significant challenges for the European Space Agency was to develop a liquid-hydrogen stage for the Ariane rocket in the 1970s. The Soviet Union did not even test a liquid-hydrogen upper stage until the mid-1980s. The Russians are now designing their Angara launch vehicle family with liquid-hydrogen upper stages.

Lack of Soviet liquid-hydrogen technology proved a serious handicap in the race of the two superpowers to the Moon.[4] Taming liquid hydrogen is one of the significant technical achievements of twentieth-century American rocketry.

From our study of Centaur history, we have identified three themes. The first is Centaur's remarkable survival in the face of three decades of attempts to cancel the program. The acceptance of liquid hydrogen as a rocket fuel demonstrates the truism that success or failure of an innovation never depends on technology alone. Thomas Hughes, a prominent historian of technology, has argued that large-scale technology organizations, whether public or private, often stifle innovation. But innovations sometimes win acceptance for reasons that are external to the technology itself.[5] Other historians have expanded on this theme, contending that technology has more to do with people, their values, and the external challenges they face than with the intrinsic superiority of one system over another.[6] We argue that liquid hydrogen as a rocket fuel would never have been adopted by NASA without the strong advocacy of Abe Silverstein and others who came out of a background in liquid-fuels research at Lewis Research Center.

Initially, during the development phase of Centaur at Marshall Space Flight Center (discussed in the first three chapters by Virginia Dawson), problems that put Centaur's survival in jeopardy included the novelty of its fuel, its controversial design, and poor management of the program. After the first test of the rocket failed, the program narrowly missed cancellation. The transfer of the Centaur program in 1962 to Lewis Research Center in Cleveland, Ohio, represented a decision by Headquarters to keep innovative aerospace companies like General Dynamics in business. At the same time, former Vanguard personnel at Cape Canaveral took over what was then called "unmanned launch operations," allowing them to develop independently of the von Braun team that managed the Apollo program on the other side of the Banana River.

In the 1970s, the proven technology of Centaur was almost cast aside during the second phase of its history when new social and organizational priorities dictated a change in policy. After a new reusable Space Transportation System (STS) received funding in 1972, expendable launch vehicles were considered obsolete. Ironically, the upper stage's most significant contributions to NASA occurred while the nation awaited the Shuttle. Centaur's service as an upper stage for Atlas-Centaur and Titan-Centaur planetary missions is discussed in chapters 4 and 5 by Mark Bowles.

[4] This point is made at various points by Asif A. Siddiqi, *Challenge to Apollo: The Soviet Union and the Space Race, 1945–1974* (Washington, DC: NASA SP-2000-4408).

[5] Thomas P. Hughes, *American Genesis: A Century of Invention and Technological Enthusiasm* (London: Penguin Books, 1989) and *Networks of Power: Electrification in Western Society, 1880–1930,* (Baltimore, MD: The Johns Hopkins University Press, 1983).

[6] See essays by Wiebe Bijker, Thomas P. Hughes, and Trevor J. Pinch, eds., *The Social Construction of Technological Systems: New Directions in the Sociology and History of Technology* (Cambridge, MA: MIT Press, 1987).

Shortly before the Shuttle began flying, the European Ariane rocket became available for launching communications satellites, ending a monopoly NASA had enjoyed since the 1960s. NASA recognized that in order for the Space Shuttle to become the all-purpose vehicle envisioned by NASA, the Shuttle required an upper stage to replicate the important job of positioning satellites and spacecraft (already performed by Atlas-Centaur). The modification of Centaur to fit in the Shuttle's cargo bay won a reprieve for the Centaur program in the 1980s. However, objections to the radical nature of its design and liquid-hydrogen fuel threatened to scuttle the program.

The loss of *Challenger* in 1986 proved another turning point in Centaur's tortuous career. Chapter 8, by Virginia Dawson, describes how Atlas-Centaur again rose like a phoenix from the ashes of *Challenger*. With NASA no longer controlling its fate, Centaur was reborn in the Atlas as a commercial launch vehicle. Commercialization of Atlas-Centaur in the 1990s provides an example of technology transfer from the government to the private sector of the economy—a policy promoted since NASA's founding. It also challenges the theory that without the stimulus of the government to pay for innovation, companies like General Dynamics (later acquired by Martin Marietta and now part of Lockheed Martin) would have no incentive to develop new technology. The Atlas rocket now faces the challenge of a highly competitive international marketplace and declining demand for communications satellite launches.

Another theme, closely related to the first, is how NASA's changing tolerance for risk influenced Centaur. Howard McCurdy pointed out in his insightful study of NASA culture that risk tolerance diminished as NASA matured.[7] Yet even in the 1960s, most of NASA's technical decisions focused on minimizing risk. For example, Administrator T. Keith Glennan had pressed Atlas into service for the Mercury program because it was a "known technology." He thought the Agency should use already-developed missiles for launch vehicles, "extending the state of the art as little as necessary."[8] Centaur was an anomaly because of its novel fuel. Glennan strongly supported its development because liquid hydrogen promised a leap in performance urgently needed to trump the Russians in space. Glennan's successor, James Webb, also became an advocate of Centaur, willing to stand up to critics like von Braun and powerful members of the U.S. House of Representatives Committee on Science and Astronautics, which controlled NASA's purse strings.

Tolerance for risk enters the Centaur story again in the 1980s. Chapters 6 and 7, by Mark Bowles, present a critical examination of one of the most interesting episodes in NASA launch vehicle history. The Shuttle/Centaur saga shows that the fate of technology can often rest upon an unplanned-for contingency—the level of social acceptance of risk. Despite the enormous advantages of sending the Galileo space probe to Jupiter on Shuttle/Centaur, the program encountered strong resistance from engineers at Johnson Space Center and the astro-

---

[7] Howard E. McCurdy, *Inside NASA: High Technology and Organizational Change in the U.S. Space Program* (Baltimore: The Johns Hopkins University Press, 1993), 150.

[8] J. D. Hunley, ed., *The Birth of NASA: The Diary of T. Keith Glennan* (Washington, DC: NASA SP-4105, 1993), 13.

naut corps located there. The astronauts thought that a liquid-hydrogen rocket in the Shuttle's cargo bay put their lives at risk, despite the fact that the main engine of the Shuttle burned liquid hydrogen carried in a large external tank. Johnson Space Center engineers also questioned the capability of the Lewis Centaur team to safely integrate Shuttle/Centaur into "their" vehicle. The *Challenger* disaster forced NASA to redefine the Agency's risk limits and led to Shuttle/Centaur's cancellation.

Risk also entered the commercialization story when General Dynamics saved Atlas-Centaur by pledging company funds to underwrite its manufacture and marketing. Commercialization ended the company's dependence on the cost-plus-fixed-fee government contract. General Dynamics/Astronautics staked the future of the company on its ability to upgrade its rocket and launch pad and sell launch services to the communications satellite industry. The gamble paid off, but–ironically–not for General Dynamics, but for its former rival and new owner, Martin Marietta.

A third theme of the book is the collaboration between government engineers and their industry counterparts. Few books, with the exception of Joan Lisa Bromberg's *NASA and the Space Industry*, have explored this relationship. Bromberg points out that Glennan did not approve of allowing the government to control the creative aspects of design "while farming out to industry only the repetitive and straight production items."[9] His successor, James Webb, had also steered the Agency away from developing an exclusively in-house technical capability. He envisioned government-industry collaboration in the management of the large-scale technology required for the space program. "These birds are going to fly," Webb told President Kennedy in 1962, "not by what you put on the schedule or the amount of money you put in it, but the way this thing [NASA] is run." He thought Centaur could "validate the capacity of the government to run a program like this in partnership with industry."[10] This was a departure from the top-down in-house Army model of working with contractors favored by Wernher von Braun. The relationship between General Dynamics and NASA under Marshall Space Flight Center (MSFC) management had been adversarial, but after the transfer to Cleveland, the relationship gradually became more collegial. Members of the NASA-industry Centaur team believed that they were on the front line of the Cold War and that Centaur held one of the keys to winning back the country's lost prestige. Lewis engineers demanded that contractors meet specifications, but they also worked closely with their industry counterparts to improve the vehicle. Despite the continual threat of cancellation, the NASA-industry team introduced a new Teledyne avionics system and developed a revolutionary steering program to adjust for upper atmosphere winds encountered immediately after liftoff.

---

[9] Quoted from Glennan's diary by Joan Lisa Bromberg, *NASA and the Space Industry* (Baltimore: The Johns Hopkins University Press, 1999), 40.

[10] Transcript of Presidential Meeting, 20 November 1962, 45.

As NASA matured, the government-industry partnership became the foundation for Centaur's commercialization, described in chapter 8. Technology transfer from the government to the private sector of the economy was realized in the 1990s. No longer held back by the government from improving Atlas-Centaur, General Dynamics made its launch vehicle comparable to the European Ariane rocket in power and versatility. Ironically, although most of the upgrades to the vehicle were made to the Centaur upper stage, the entire vehicle is now referred to simply as the Atlas. Thus Centaur, still a vital part of the vehicle, has become invisible.

NASA Lewis (now Glenn) Research Center engineers and procurement officers helped to unravel the government's complex contractual relationship with General Dynamics. Commercialization liberated General Dynamics from a culture of government dependency. Today, although human bonds formed over three decades are still strong, NASA Glenn no longer manages Centaur. The limited oversight the government retains for NASA missions is currently carried out by Kennedy Space Center.

Few secondary sources recognize the role Centaur played in the controversies surrounding the acceptance of liquid hydrogen and its positive contributions to the unmanned space program. For example, in *On Mars*, Edward and Linda Ezell dismiss Centaur as "a genuine troublemaker" because of the delays it caused for the Surveyor and Mariner programs.[11] In Clayton R. Koppes's critical study of the Jet Propulsion Laboratory's early unpiloted projects, *JPL and the American Space Program*, Atlas-Agena and Atlas-Centaur receive scant notice,[12] while a major new encyclopedia of America's greatest space programs leaves out the Centaur program altogether.[13] An exception is the late John Sloop, whose engagement with high-energy liquid rocket propellants in the 1950s shaped his career at the Lewis Flight Propulsion Laboratory in Cleveland, one of three laboratories for the NACA. After NASA's formation, he became Abe Silverstein's technical assistant at Headquarters. Sloop recognized the pivotal role of liquid hydrogen in the American space program, and upon retirement, he published *Liquid Hydrogen as a Propulsion Fuel, 1945–1959*, a detailed and insightful study that concludes in 1959, before NASA took over Centaur program management from the Air Force.[14] His interviews (now part of the NASA History Collection) with former General Dynamics employees, notably Krafft Ehricke and Deane Davis, as well as other papers in the NASA Historical Collection, provided an indispensable starting point for our study. Another source that provided essential background in understanding Centaur's significance was

[11] Edward Clinton Ezell and Linda Neuman Ezell, *On Mars* (Washington, DC: NASA SP-4212, 1984), 47.

[12] Clayton R. Koppes, *JPL and the American Space Program* (New Haven: Yale University Press, 1982).

[13] Frank N. Magill and Russell R. Tobias, eds., *USA in Space*, three volumes (Pasadena, CA: Salem Press, Inc., 1996).

[14] John L. Sloop, *Liquid Hydrogen as a Propulsion Fuel 1945–1959* (Washington, DC: NASA SP-4404, 1978). A native of North Carolina, Sloop earned a B.S. in electrical engineering from the University of Michigan in 1939. Hired by the NACA in 1941, he moved to Cleveland in 1942 and remained at Lewis until his move to Washington, DC, in 1958.

Roger E. Bilstein's *Stages to Saturn*.[15] Several superb papers by Richard Martin, formerly of General Dynamics, contributed to our understanding of the unique structure of the Centaur.[16] Articles by G. R. Richards and Joel W. Powell, and Joseph Green and Fuller C. Jones are among the few specifically focused on Centaur's origins and early history.[17] Our understanding of the development of Pratt & Whitney's RL10 engine was enhanced by work of Joel Tucker and Dick Mulready.[18] Finally, the work by John Krige on the European Space Agency provided valuable background for our chapter on commercialization.[19]

One of the themes in Virginia Dawson's earlier book, *Engines and Innovation*,[20] a history of the NASA Lewis Research Center, was the tension between the research side of the laboratory and the real-time demands of managing the Agena and Centaur missions for NASA. The management of Centaur is arguably Lewis Research Center's most important contribution to the space program. In this book, we emphasize how the research culture and test facilities of the Center materially contributed to Centaur development. If, as some of the reviewers of our manuscript from General Dynamics have commented, we may have overstated the Lewis role, we can only offer the hope that a well-documented history of General Dynamics/Astronautics[21] can redress any imbalance that we may have created. Although we were fortunate to be able to interview some former General Dynamics employees, we were disappointed that very few company documents can be found in publicly accessible archives.

This history of Centaur is the result of an association that began nine years ago when NASA history first brought the authors together. We collaborated on an essay about NASA's Advanced

---

[15] Roger E. Bilstein, Stages to Saturn (Washington, DC: NASA SP-4206, 1980).

[16] See Richard E. Martin, "The Atlas and Centaur 'Steel Balloon' Tanks: A Legacy of Karel Bossart," 40th Congress of the International Astronautical Federation, Malaga, Spain, 8–14 October 1989, IAA-89-738, Cot 7-13m, reprint; "A Brief History of the Atlas Rocket Vehicle, Part III," *Quest—The History of Spaceflight Quarterly* 8 (2001): 48; "Atlas II and IIA Analyses and Environments Validation," *Acta Astronautica* 35 (1995): 771–791.

[17] G. R. Richards and Joel W. Powell, "The Centaur Vehicle," *Journal of the British Interplanetary Society* 42 (1989): 99–120; Joseph Green and Fuller C. Jones, "The Bugs That Live at -423°," *Analog Science Fiction/Science Fact* 80 (1968): 8–41.

[18] See Joel E. Tucker, "The History of the RL 10 Upper-Stage Rocket Engine, 1956–1980," in *History of Liquid Rocket Engine Development in the United States, 1955–1980*, ed. Stephen E. Doyle, AAS History Series (San Diego: AAS, 1992), vol. 13, 123–151. Also, Dick Mulready, *Advanced Engine Development at Pratt & Whitney: The Inside Story of Eight Special Projects, 1946–1971* (Warrrendale, PA: Society of Automotive Engineers, Inc., 2001), chapters 3–4.

[19] See J. Krige, A. Russo, and L. Sebesta, *A History of the European Space Agency, 1958–1987* (ESA SP-1235 April 2000), vol. 2.

[20] Virginia P. Dawson, *Engines and Innovation: Lewis Laboratory and American Propulsion Technology* (Washington, DC: NASA SP-4306, 1991).

[21] A company history by Bill Yenne, *Into the Sunset: The Convair Story* (Lyme, CT: Greenwich Publishing Group, 1995), could be used as a starting point.

Turboprop Project in a book on Collier trophy winners.[22] Since that time, we have coauthored a variety of projects that have come to us through our company, History Enterprises, Inc. We have enjoyed ferreting out the technology involved in developing and managing the Centaur rocket and have tried to achieve a balance among the technology, the scientific goals of the missions themselves, and the political aspects of the program. Our first essays on Centaur were published in *To Reach the High Frontier*, edited by Roger D. Launius and Dennis R. Jenkins.[23]

We bring very different perspectives to NASA history. Mark Bowles grew up reveling in the achievements of the planetary probes of the 1970s. As a boy, he shared with his father the deep sense of awe in the mysterious dark universe surrounding Earth's tiny blue oasis of life. He recalls how they would gaze up at the Moon during the Apollo missions and imagine the activities of the astronauts.

Virginia Dawson's engagement with NASA history came through the encouragement she received from the late Professor Melvin Kranzberg, founder of the field of the history of technology, who served as chair of the NASA History Advisory Committee for many years. In his foreword to *This New Ocean: A History of Project Mercury*, he pointed out that the legislation creating the National Aeronautics and Space Administration in 1958 included not only the charge to expand "human knowledge of phenomena in the atmosphere and space," but also to provide for the widest possible dissemination of information about its activities. "NASA wisely interpreted this mandate to include responsibility for documenting the epochal progress of which it is the focus," he wrote. "The result has been the development of a historical program by NASA as unprecedented as the task of extending man's mobility beyond his planet."[24] Dr. Kranzberg strongly believed that professional historians working outside academe could make valuable contributions to the history of twentieth-century institutions. He was particularly impressed that NASA gave authors unrestricted access to participants and unclassified sources and encouraged them to examine Agency history critically.

We enjoyed the assistance and encouragement of many current and former NASA people. First, we would especially like to thank Joe Nieberding, who enthusiastically took us under his wing in the early stages of the book and pushed the project along. He opened doors to interviews with people at Glenn Research Center, Kennedy Space Center, the Jet Propulsion Laboratory, and Lockheed Martin, as well as former employees of General Dynamics in San Diego. We would also like to thank Meyer Reshotko for helping us get started on this project. J. Cary Nettles, Roger

[22] Mark D. Bowles and Virginia P. Dawson, "The Advanced Turboprop Project: Radical Innovation in a Conservative Environment," in *From Engineering Science to Big Science: The NACA and NASA Collier Trophy Research Project Winners*, ed. Pam Mack (Washington, DC: NASA SP-4219, 1998), 321–343.

[23] Virginia P. Dawson, "Taming Liquid Hydrogen: The Centaur Saga," 334–356, and Mark D. Bowles, "Eclipsed by Tragedy: The Fated Mating of the Shuttle and Centaur," 415–442, *To Reach the High Frontier: A History of U.S. Launch Vehicles*, ed. Roger D. Launius and Dennis Jenkins (Lexington: The University Press of Kentucky, 2002).

[24] Loyd S. Swenson, Jr., James M. Grimwood, and Charles C. Alexander, *This New Ocean: A History of Project Mercury* (Washington, DC: NASA SP-4201, 1966), V.

Lynch, and Richard Martin gave us copies of key photos. We are greatly indebted to the people who patiently reviewed our draft of the manuscript. These included Joe Nieberding, Frank Spurlock, Richard Martin, Larry Ross, Edward Bock, and John Neilon. History colleagues John Hunley and Dwayne Day not only read our manuscript with extraordinary care and insight, but also shared with us documents and articles important to our story. Alan Lovelace, Marty Winkler, Ron Everett, Cary Nettles, Art Zimmerman, Len Perry, and Del Tischler reviewed individual chapters. All their comments proved exceedingly helpful. However, we take responsibility for our interpretation of events and any errors of fact that might have gone uncorrected. We are also very grateful to the many people we interviewed, either by telephone or in person. They are listed at the end of the book. Our understanding was also enhanced less formally by many others too numerous to cite here. Many of the people we interviewed spoke of the extraordinary dedication of Steve Szabo. Because of his untimely death, we were not able to interview him, but we hope that we have captured some of the Centaur team spirit that he helped to create.

As colleagues in NASA history will appreciate, finding documentation for government-funded studies is always a challenge. For this history, we were fortunate to have the assistance of Kevin Coleman, in charge of the history program and records management at Glenn; he, along with Deborah Demaline, proved to be a judicious and supportive ally. We used Glenn records stored in the famous World War II bunkers at Plum Brook and at the National Archives and Record Center in Chicago. Bonita Smith made certain we had access to Centaur material in the NASA Glenn Archives. Janice Nay and Lynn Patterson provided excellent transcripts of our oral interviews. Robert Arrighi of History Enterprises assisted with research.

During the writing of *Engines and Innovation*, Virginia Dawson and historian of science colleague Craig Waff discovered a trove of records of the Launch Vehicles Division stored in the vault of the Development Engineering Building at Glenn. In looking for Centaur records more than ten years later, we found that some boxes containing correspondence fortuitously were left behind in the move to Kennedy Space Center. Mark Bowles was able to use some of this material to document the Shuttle/Centaur story. The contents of these boxes, catalogued by Galen Wilson, are referred to as NASA GRC Records to distinguish them from the NASA GRC Archives. The ELV Resource Library, managed by Boeing at Kennedy Space Center, loaned us a selected set of historical Lewis documents. Robert Bradley made available Krafft Ehricke's early Air Force reports, now located in the archives of the San Diego Aerospace Museum. We also used Centaur records in the NASA Historical Reference Collection in the NASA History Office in Washington, District of Columbia, many of which were assembled through the efforts of the late John Sloop. We are grateful for the assistance and encouragement we received from Roger Launius, Steve Dick, Steve Garber, Louise Alstork, and Jane Odom in the NASA History Office and Mike Wright at Marshall Space Flight Center.

Last, but certainly not least, many thanks are due to the professional graphic designers, editors, and print specialists who made this book a physical reality. In the NASA Headquarters Printing and Design Office, Douglas Ortiz, Joel Vendette, and James Gitlin expertly handled the layout of the book; Lisa Jirousek carefully edited the book; and Jeffrey McLean and James Penny took care of the printing process.

Chapter 1

Centaur's Origins in Atlas

"Going back to the old ideas of Oberth, I said I have a relatively dense first stage. And it so happens that the second stage, because it is less dense, fits just beautifully on the first stage. Now all we have to do is remove the neck from Atlas and make it cylindrical all the way and we have a 10-foot-diameter base. For a second stage, that's just beautiful."

—Krafft Ehricke, Centaur's designer

In early 1956, the development of Atlas preoccupied Krafft Ehricke. Known for his enormous capacity for work, he spent 18-hour days driven by the urgency of producing the country's first intercontinental ballistic missile (ICBM). At the same time, Ehricke envisioned more peaceful applications for Atlas. Why not add a second stage to produce a two-stage rocket capable of placing communications or weather satellites into orbit, sending instrumented space probes to the Moon and the planets, or installing an orbiting platform around Earth? He even calculated how the Atlas itself could become a satellite when launched into low-Earth orbit.

This was the kind of fantasy that Ehricke had entertained ever since his youth in Weimar, Germany. Mesmerized by the 1929 Fritz Lang film *Frau im Mond (Woman in the Moon)*, which he saw more than a dozen times, he said, "I felt, 'My God, it really must be possible to get to the Moon,' which for an 11-year-old boy is a kind of revelation."[1] The son of two dentists, Ehricke tinkered with their dental apparatus and chemical compounds while he steeped himself in the ideas of the early rocket pioneers. He recalled the inspiration for Centaur: "Going back to the old ideas of Oberth, I said I have a relatively dense first stage. And it so happens that the second stage,

---

[1] Interview with Krafft Ehricke by John Sloop, 26 April 1974, NASA Historical Reference Collection, NASA History Office, Washington, DC. On Ehricke's biography and ideas, see Marsha Freeman, *How We Got to the Moon: the Story of the German Space Pioneers* (Washington, DC: 21st Century Associates, 1993), 292–339; and "Krafft Ehricke's Extraterrestrial Imperative: A Memoir," in *History of Rocketry and Astronautics*, eds. Donald C. Elder and Christophe Rothmund (San Diego: American Astronautical Society, 2001), AAS History Series, vol. 23, 163–222. See also Al Vinzant, "History of Centaur," *Exclusive* (GD Space Systems Division Newsletter), August 1988; and Daniel A. Heald, "LH2 Technology Was Pioneered on Centaur 30 Years Ago," 43rd Congress of the International Astronautical Federation, August 1992, Washington, DC.

Krafft Ehricke, Centaur's designer and first program director (left), discusses advanced spacecraft designs with James Dempsey, president of General Dynamics/Astronautics, early 1960s. (Courtesy of Lockheed Martin)

because it is less dense, fits just beautifully on the first stage. Now all we have to do is remove the neck from Atlas and make it cylindrical all the way and we have a 10-foot diameter base. For a second stage, that's just beautiful."[2] Because the two-stage rocket he envisioned for spaceflight needed a very powerful upper stage, he considered high-energy propellant combinations like liquid hydrogen with fluorine as the oxidizer, hydrazine with fluorine, methane with ozone, and liquid hydrogen with liquid oxygen.[3] These studies smacked of science fiction in 1956, and Ehricke expected James Dempsey, his boss at the Convair Division of General Dynamics, to put an end to them. Instead, Dempsey urged him to continue.[4]

    Ehricke championed the idea of space travel. He chaired the Space Flight Committee for the American Rocket Society, which comprised some of the nation's most prominent missile

[2] Interview with Krafft Ehricke by John Sloop, 26 April 1974, NASA Historical Reference Collection.

[3] Al Vinzant, "History of Centaur."

[4] Shirley Thomas, *Men of Space*, vol. 1 (Philadelphia: Chilton Company, 1960), 2.

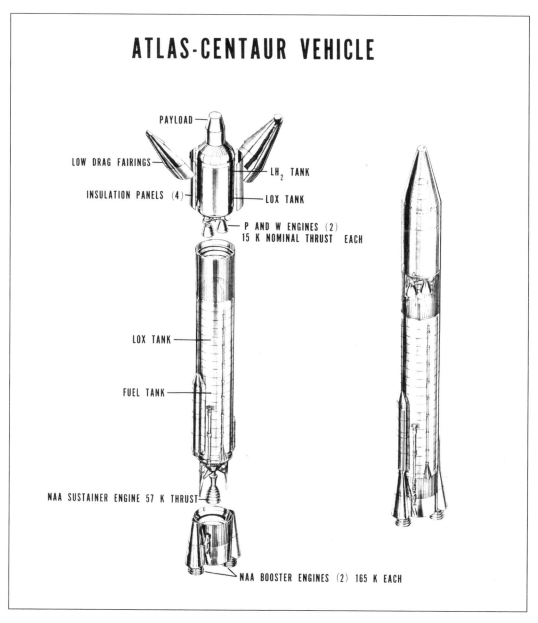

ATLAS-CENTAUR VEHICLE

PAYLOAD

LOW DRAG FAIRINGS

LH$_2$ TANK

INSULATION PANELS (4)

LOX TANK

P AND W ENGINES (2)
15 K NOMINAL THRUST EACH

LOX TANK

FUEL TANK

NAA SUSTAINER ENGINE 57 K THRUST

NAA BOOSTER ENGINES (2) 165 K EACH

Diagram: Atlas-Centaur. (NASA E2283)

designers, including Karel "Charlie" Bossart of Convair, Wernher von Braun of the Army Ballistic Missile Agency, Milton Rosen of the Naval Research Laboratory, and Hubertus Strughold of Randolph Air Force Base. Shortly before the Soviet Union launched Sputnik, the world's first artificial satellite, the committee finished a report that they then submitted to President Eisenhower less than two weeks after Sputnik's launch. The report proposed a national space agency to benefit all mankind.[5] Ehricke thought the new space age would usher in a new age of discovery and international cooperation. He envisioned space travel "as a serious, practical and worthwhile effort—not at some future time, but right now, in this century and in this age of ours." He thought space travel appealed to man's highest aspirations and would promote international peace and goodwill.[6]

In response to Sputnik, James Dempsey asked Ehricke to proceed with a design for an upper stage for the Atlas missile, transforming the weapon into a two-stage spacecraft for travel to the outer reaches of the solar system. He was encouraged to move offsite with an elite team of the company's engineers that included Atlas veterans Charlie Bossart, William Patterson, Howard Dunholter, Frank Dore, William Radcliffe, James Crooks, and Hans Friedrich, a German expatriate who had designed the autopilot for the German V-2 rocket. After debating the merits of various propellant combinations, the group agreed that liquid hydrogen with the oxidizer liquid oxygen was the best propellant for the high-energy upper stage Ehricke had in mind. Ehricke may have named Centaur for the nearest star after the Sun, the Alpha Centauri, which might one day be approached by the vehicle he envisioned.[7] It has also been speculated that Ehricke chose the name because the mythic man-beast with the body of a horse and the torso and head of a man reflected the hybrid nature of the Atlas-Centaur combination, with Centaur carrying the guidance and control for both vehicles.

Despite the challenges of producing, storing, and handling liquid hydrogen, its distinct advantages as a rocket propellant presented the rocket team at General Dynamics with a tantalizing opportunity at a time when U.S. launch vehicle capability was well below that of the Soviet Union. Burning liquid hydrogen with liquid oxygen produces the highest specific impulse or thrust of any known rocket fuel. Theoretically, a liquid-hydrogen-powered rocket can lift approximately 35 to 40 percent more payload per pound of liftoff weight than conventional rocket fuels.[8]

Hydrogen had remained a laboratory curiosity until the late nineteenth century, when scientists driven by the promise of industrial applications for gases began to investigate how to convert gases to liquids. The British scientist Michael Faraday succeeded in liquefying chlorine, ammonia, and carbon dioxide. Oxygen and hydrogen proved more difficult to liquefy because they are cryogenic—gases that can be converted to liquids only at extremely low temperatures. In the 1880s, a method was found for liquefying oxygen. Then, in 1898, Sir James Dewar succeeded in liquefying hydrogen for the first

---

[5] "ARS Urges National Space Flight Program," *Astronautics* 2 (December 1957): 18–28.

[6] Krafft A. Ehricke, "The Anthropology of Astronautics," *Astronautics* 2 (November 1957) 28.

[7] The Alpha Centauri suggestion is based on information from Harold Dunholter, former chief development engineer for General Dynamics. See Helen T. Wells, Susan H. Whiteley, and Carrie E. Karegeannes, *Origin of NASA Names* (NASA SP-4402, 1976), 11–12.

[8] Centaur, a pamphlet prepared by the Convair Aerospace Division of General Dynamics, dated May 1973, provided the 35-percent figure. Generally, NASA press releases use 40 percent. Centaur files, NASA Glenn History Office.

time. Dewar described the principle of regenerative cooling, a method that takes advantage of the extreme cold of liquid hydrogen to cool hot metal surfaces during combustion. Dewar also invented vacuum containers for safely storing liquid hydrogen.[9]

Liquid oxygen found applications in the 1920s as an oxidizer for alcohol or kerosene fuel, but early rocket pioneers like Konstantin Tsiolkovskiy and Robert Goddard avoided liquid-hydrogen fuel because of its well-known dangers.[10] Hermann Oberth had calculated that liquid hydrogen was especially appropriate for an upper stage because in the near-vacuum of space, less pressure is required to keep the sides of the rocket rigid.[11] However, like Goddard and Tsiolkovskiy, Oberth had avoided testing an oxygen/hydrogen rocket, presumably because of hydrogen's volatility. The tragic explosion of the hydrogen-gas-filled Hindenberg dirigible in 1937 seems to have reinforced the general prejudice against hydrogen.

Although Ehricke understood the theoretical advantages of liquid hydrogen, he had grossly underestimated the difficulties of developing a liquid-hydrogen rocket. A rocket enthusiast during his student years in Berlin, he had filed two rocket patents, helped organize the German Society for Space Research in 1938, and written a series of articles for the Society's journal *Space*.[12] But he had avoided testing a liquid-hydrogen/liquid-oxygen rocket because of the fear that the mixture might explode prematurely.[13] Graduate studies in nuclear physics at Humboldt University under Werner Heisenberg were interrupted by the war. Ehricke served at the Russian front as a tank driver before being recruited for work on the V-2 at Peenemünde. There he came under the spell of Walter Thiel, Wernher von Braun's rocket engine expert. Thiel inspired Ehricke to imagine rocket engines with millions of pounds of thrust. Learning of Heisenberg's attempts to build a nuclear reactor, Ehricke and Thiel dreamed of nuclear rocket propulsion with hydrogen as the working fluid.[14]

Thiel gave careful attention to the possibility of developing a liquid-hydrogen rocket. In a significant memorandum written in 1937, Thiel made a survey of existing knowledge of the

---

[9] John D. Clark, *Ignition! An Informal History of Liquid Rocket Propellants* (New Brunswick, NJ: Rutgers University Press, 1972), 4–5.

[10] See *Exploring the Unknown: Selected Documents in the History of the U.S. Civil Space Program*, vol. I (NASA SP-4218, 1995), 67–68; John Sloop, *Liquid Hydrogen as a Propulsion Fuel*, 1945–1959 (Washington, DC: NASA SP-4404, 1978), 259.

[11] In discussing hydrogen fuel, Hermann Oberth wrote, "The thinner the outside air, the smaller are the forces which tend to cause the rocket to buckle or collapse and the smaller is the inside pressure necessary for rigid filling." See *Ways to Spaceflight, tr. Wege zur Raumschiffahrt*, originally published in 1929 (Washington, DC: NASA TT F-622, 1972), 343.

[12] Marsha Freeman, "Krafft Ehricke's Extraterrestrial Imperative," 164.

[13] Interview with Krafft Ehricke by John Sloop, 26 April 1974, NASA Historical Reference Collection.

[14] Interview with Krafft Ehricke by John Sloop, 26 April 1974, NASA Historical Reference Collection.

"practical possibilities" for improving liquid rockets.[15] Thiel drew attention to the very large energy gain that could be obtained from using high-energy propellants such as liquid hydrogen/liquid oxygen and stated that a research effort "appears definitely rewarding." At the same time, he tempered his enthusiasm for liquid hydrogen by pointing out the "strong obstacles" to the development of a liquid-hydrogen rocket, such as the need for insulated tanks and ducts to compensate for the extremely low temperatures of liquid hydrogen/liquid oxygen. He pointed out that very little was known about the behavior of metals at extremely low temperatures, particularly the permeability of light metals to liquid hydrogen. This influential rocket expert was killed in an Allied bombing raid on Peenemünde, but his ideas probably took root in the fertile mind of Krafft Ehricke. After Peenemünde, the possibilities of liquid hydrogen continued to intrigue him.

In 1945, Ehricke managed to obtain a contract from the Department of the Army and joined the German rocket group at White Sands, New Mexico. Between 1947 and 1950, he worked on ramjet and rocket systems at Fort Bliss, Texas. Because there was little information to guide rocket designers in the selection of propellants, the *Journal of the American Rocket Society* devoted an entire issue to the subject in 1947. A *Journal* paper by Maurice Zucrow argued that, in addition to thrust, propellant density must also be considered in assessing rocket performance, since it can affect vehicle size and weight. In a chart, Zucrow showed that the liquid-oxygen/liquid-hydrogen combination produced greater thrust but required a larger tank because the low density of hydrogen took up more volume.[16]

Ehricke recalled being impressed by a paper by JPL researcher Richard Canright that urged more research on high-energy liquid propellants, especially liquid hydrogen with the oxidizer hydrazine.[17] In a 1974 interview, Ehricke said, "I had again run into hydrogen, from Tsiolkovskiy through von Hoefft to Oberth to Thiel and through my nuclear investigation. So I said, 'It's too often that it has looked good. I think we ought to do it.'" But von Braun thought they should stick with the denser propellants they already knew how to handle.[18]

---

[15] Walter Thiel, "Memorandum of the Practical Possibilities of Further Development of the Liquid Rockets and a Survey of the Tasks to be Assigned to Research," 13 March 1937, tr. D. K. Huzel, folder CT-168000-01, "Thiel, Walter," National Air and Space Museum Archives. Thanks to historian John Hunley for sending us a copy of this document.

[16] Ehricke says the paper was published in *Journal of the American Rocket Society* in 1947 and authored by Jet Propulsion Laboratory rocket researcher Richard Canright. There is no paper by Canright in the journal between 1947 and 1953. We believe that the paper referred to here is by Maurice Zucrow in the 1947 issue devoted to liquid propellants: "Liquid Propellant Rocket Power Plants," no. 72 (December 1947): 28–44.

[17] Interview with Krafft Ehricke by John Sloop, 26 April 1974, NASA Historical Reference Collection. A review of the *Engineering Index* revealed that Richard Canright published "Problems of Combustion: Liquid Propellant Rocket Motors," in *Chemical Engineering Progress* 46 (1950): 228–232. There is, however, no indication of a paper on a comparison of rocket propellants.

[18] Interview with Krafft Ehricke by John Sloop, 26 April 1974, NASA Historical Reference Collection. See also John Sloop, *Liquid Hydrogen*, 47–48.

In 1950, Ehricke moved with the German team to Huntsville, Alabama, where he joined the Guided Missile Development Group at the Redstone Arsenal. He headed the Gas Dynamics Section but chafed under von Braun's authority and his conservative approach to rocket design.[19] In a rocket, where every excess pound reduces the size of the possible payload and increases cost, Ehricke paid close attention to propellant weights. In a paper comparing propellants and working fluids for rocket propulsion, he compared the weight of propellants, their specific impulse, and their density. He argued that for ascent in a gravity field, heavy or medium propellants were preferable. However, once a staged rocket reached space, the performance of light, less dense propellants like ozone-methane, hydrogen-oxygen, and oxygen-hydrazine increased. They beat the heavier propellants because in zero gravity, their high specific impulse and low density offset the weight of the structures needed to contain them. He suggested that liquid-hydrogen upper stages needed investigation.[20]

At the end of his contract with the U.S. Army in 1952, Ehricke moved to a job at the Bell Aircraft Corporation in Buffalo under Walter Dornberger, another German expatriate who had worked on the V-2. In 1954, Ehricke joined Convair in San Diego as a design specialist in the same year that it became a division of General Dynamics. He headed the Preliminary Design and Systems Analysis group. He later served as the director of the Centaur program for General Dynamics/Astronautics from 1958 until January 1962, when he became the director of Advanced Studies.[21]

The Thin-Skinned Atlas

At Convair, Ehricke found an innovative engineering culture that had evolved under the technical direction of Charlie Bossart, an unassuming Belgian émigré. The "father of the Atlas" was born in Antwerp. He had earned a degree in mining engineering from the Free University of Brussels in 1925. During his last year at the university, an optional course in aeronautical engineering earned him a scholarship to study aeronautical engineering at MIT. After receiving a master's degree in 1927, he returned to Belgium for military service, then emigrated to the United States in 1930. He worked for Sikorsky at General Aviation in aircraft structures before moving to Downey, California, to work for Vultee Aircraft in 1941. The company became Consolidated Vultee, or Convair, in 1943.

Bossart's background in structures gave him a different perspective from which to approach rocket design. He conceived the innovative pressure-stabilized design of the Atlas while working on an early Air Force missile project in 1946 called the MX-774. Like many postwar rocket designers,

---

[19] Interview with Krafft Ehricke by John Sloop, 26 April 1974, NASA Historical Reference Collection. See also John Sloop's *Liquid Hydrogen as a Propulsion Fuel: 1945–1959*, 191–195.

[20] Krafft A. Ehricke, "Comparison of Propellants and Working Fluids for Rocket Propulsion," *American Rocket Society* 23 (1953): 287–296, 300.

[21] Biography introducing article in *Astronautics* 2 (1957): 26. See also interview with Krafft Ehricke by John Sloop, 26 April 1974, NASA Historical Reference Collection. In 1957, Convair became the Convair/Astronautics Division of General Dynamics; in 1961, the name was changed to General Dynamics/Astronautics, a Division of the General Dynamics Corporation.

Bossart's group had taken the V-2 as their starting point. However, unlike the rocket team that worked under von Braun at White Sands, they were free of Peenemünde's traditions and prejudices.

Bossart's design dispensed with the heavy rings and stringers of the V-2. Ring and stringers were metal bands connected by longitudinal metal rods that reinforced the inside walls of the rocket. This rigidity provided an extra margin of safety, but it also added unnecessary weight. The originality of the MX-774 lay in its lack of the internal buttressing. Pressure alone kept the sides of the rocket's tank rigid when inflated like a balloon with its fuel. Its stainless-steel walls were no thicker than a dime. Although it was not an entirely new idea, since Hermann Oberth in Germany had discussed its theoretical possibility, Bossart was the first to demonstrate the feasibility of a pressure-stabilized rocket design. Bossart always attributed his confidence in the thin-skinned design to his background in aircraft structures. He wrote:

> You see most of my previous experience was in structures. So the first thing I did was to decide what kind of a structure we're going to use for this missile. Well, we knew that we had to have a certain pressure in the tank to maintain the required net positive suction head for the pump. So the first question was, how thin does the skin have to be to resist that pressure? With the skin thickness thus arrived at and assuming the tank to be under pressure, how much stiffening would you have to add to make this thing capable of taking the compression and the bending anticipated in flight? And lo-and-behold, we didn't need to add any.[22]

Even after the Air Force canceled the MX-774 project in 1947, Bossart continued to push development of the rocket with missionary zeal. His tenacity paid off in the early 1950s after technical advances allowed thermonuclear warheads to be scaled down to fit inside the nose of a missile. The rocket now won the backing of two influential Air Force officials, Trevor Gardner, Assistant Secretary of the Air Force for Research and Development, and Lieutenant Colonel Bernard Schriever.[23] Concerned that the United States was in a race with the Soviet Union in ICBM development, Gardner and Schriever advocated setting up a crash program comparable in scale and funding to the Manhattan project.

---

[22] Interview with K. J. Bossart by John Sloop, 27 April 1974, NASA Historical Reference Collection, Washington, DC. For a superb article on Bossart and the balloon tank, see Richard E. Martin, "The Atlas and Centaur 'Steel Balloon' Tanks: A Legacy of Karel Bossart," reprint by General Dynamics Corp., 40th International Astronautical Congress paper, IAA-89-738, 1989.

[23] See Frank L. Winter, *Rockets into Space* (Cambridge, MA: Harvard University Press, 1990); Edmund Beard, *Developing the ICBM: A Study in Bureaucratic Politics* (NY: Columbia University Press, 1976); Jacob Neufeld, *Ballistic Missiles in the United States Air Force, 1945–1960* (Washington, DC: Office of Air Force History, 1990); John Clayton Lonnquest, "The Face of Atlas: General Bernard Schriever and the Development of the Atlas Intercontinental Ballistic Missile, 1953–1960," Ph.D. dissertation, Duke University, 1996. See also Davis Dyer, "Necessity Is the Mother of Convention: Developing the ICBM, 1954–1958," *Business and Economic History* 22 (1993): 194–209; and *TRW: Pioneering Technology and Innovation since 1900* (Boston: Harvard Business School Press, 1998), 167–194. A review of Atlas development to the present can be found in Dennis R. Jenkins, "Stage-and-a-Half, The Atlas Launch Vehicle," in *To Reach the High Frontier: A History of U.S. Launch Vehicles*, eds. Roger D. Launius and Dennis R. Jenkins (Lexington: The University of Kentucky Press, 2002), 70–102.

Charlie Bossart, designer of Atlas, holds up a sketch of the missile, early 1960s. (Courtesy of Lockheed Martin)

A blue ribbon committee of scientific, industrial, and academic leaders headed by Princeton physicist John von Neumann endorsed the development of a long-range ballistic missile. The committee, unconvinced that the Air Force had sufficient technical depth to manage the project, recommended that the Ramo-Wooldridge Corporation assume the responsibility for technical direction of the program, reporting directly to Schriever. Prior to approval of the design in January 1955, Ramo-Wooldridge put pressure on Bossart to abandon the pressure-stabilized design because of the tendency of the bulkhead to collapse if the tanks were not pressurized correctly. To Bossart's relief, Ramo-Wooldridge agreed that development was too far along to change to the type of reinforced structure favored by most rocket designers.[24]

Engineers at Convair were intensely proud of the originality of the Atlas design, which they liked to call "the free world's first ICBM." However, not all rocket experts celebrated the success of Atlas. The group at Huntsville "looked askance at such lightweight structural innovations as Bossart's thin-wall, pressurized tanks for the Atlas ICBM, which they jokingly referred to as 'blimp' or 'inflated competition.'"[25]

During Atlas development, the company built launch complexes at Cape Canaveral and static firing test facilities at Sycamore Canyon, northeast of San Diego. It also carried out a research program at Edwards Air Force Base near Boron, California. Originally, the group working on Atlas was based in the Lindbergh Field plant located next to the San Diego airport. As it grew from about 20 employees in 1951 to 200 in 1955, the company began construction on a large plant about ten miles north of downtown San Diego at Kearny Mesa. Manufacture of the technically demanding stainless steel tanks for Atlas (and later the even more stringent requirements for the Centaur tanks) initially was located at Kearny Mesa, but several years later was moved to Plant 19, owned by the Air Force, near the Lindbergh Field plant.

In 1957, to strengthen management, General Dynamics Convair set up a separate Astronautics Division under James Dempsey, an Alabamian with a flair for handling both administrative and technical problems. A graduate of West Point, Dempsey had served in World War II. After the war, he earned a master's degree in aeronautical engineering at the Air Force Guided Missile School at the University of Michigan, served at the Pentagon and Patrick Air Force Base near Cape Canaveral, then accepted a position with the Convair Corporation in 1953.

---

[24] Interview with K. J. Bossart by John Sloop, 27 April 1974, NASA Historical Reference Collection. On the Air Force approach to contracting, see also Joan Lisa Bromberg's chapter, "Legacies," in *NASA and the Space Industry* (Baltimore: The Johns Hopkins University Press, 1999), 16–44.

[25] John Sloop, *Liquid Hydrogen*, 208. See also G. R. Richards and Joel W. Powell, "The Centaur Vehicle," *Journal of the British Interplanetary Society* 42 (1989): 99–120. Missile concepts pioneered on Atlas included the "stage and a half" launch concept. After firing all three engines at liftoff, two were jettisoned after 140 seconds, leaving the main engine to power the vehicle toward its target. Bossart introduced the idea of a separable nose cone to save weight and facilitate reentry. Other innovations included a radio-inertial guidance system, which increased the accuracy of the booster, swiveling rocket motors, and small vernier motors for greater accuracy and velocity control.

With national security riding on the successful development of the Atlas, pressure on the Astronautics Division was intense. Atlas failed during the first test on 11 June 1957, although the controversial balloon structure was validated during its dramatic tumble earthward. Despite the extreme lateral stresses placed on the tank, it did not break up until it was destroyed by the range safety officer. The launch of Sputnik the following October struck a devastating blow to the country's sense of security.

Looking back on NASA's early years, Administrator T. Keith Glennan admitted, "In truth, we lacked a rocket-powered launch vehicle that could come anywhere near the one possessed by the Soviets. And it would take years to achieve such a system, no matter how much money we spent."[26] Without vehicles with adequate thrust, the space program could achieve none of its goals, which included an ambitious program of planetary exploration. Moreover, the inability to launch even a relatively light satellite seemed to indicate that the United States lacked a sufficiently powerful missile to retaliate in the event of attack by the Soviet Union.

In the politically charged post-Sputnik atmosphere, ARPA Director Roy Johnson asked Jim Dempsey whether the company was capable of launching anything resembling a spacecraft. Recalling Ehricke's early studies of spacecraft, Dempsey bravely replied, "Sure, the whole Atlas."[27] This led to a top-secret crash program called Project SCORE (Signal Communication by Orbiting Relay Equipment). Assured that the Atlas would not fail, Glennan convinced President Eisenhower to use Project SCORE as a not-so-subtle proxy to demonstrate the country's military prowess. Glennan noted in his diary: "The recent successes of the Atlas launches would permit him [the President] to portray dramatically—possibly by television or a demonstration for the press—the effectiveness of the Atlas as a nuclear warhead carrier."[28] The Atlas, launched from Cape Canaveral on 18 December 1958 with a pretaped Christmas message of peace on Earth from President Eisenhower, carried the largest American payload placed in orbit up to that time.[29]

The increasing reliability of the Atlas encouraged NASA officials to press it into service for the Mercury program, on the theory that to catch up with the Soviet Union, NASA should "use known technologies, extending the state of the art as little as necessary."[30] Atlas would launch Mercury astronauts John Glenn, M. Scott Carpenter, Walter Schirra, and L. Gordon Cooper, but without an upper stage, it was consigned to low-Earth orbit.

---

[26] See Glennan's comments on this issue in *The Birth of NASA: The Diary of T. Keith Glennan*, ed. J. D. Hunley (Washington, DC: NASA SP-4105, 1993), 23.

[27] William H. Patterson, "The Evolution of a New Technology Concept in the USA (America's First ICBM—the Atlas)," typescript, no date, 19. Copy courtesy of Charles Wilson.

[28] J. D. Hunley, ed., *The Birth of NASA*, 24.

[29] Deane Davis, "The Talking Satellite: A Reminiscence of Project Score," *Journal of the British Interplanetary Society* 52 (1999): 239–259. See also John L. Chapman, *Atlas: The Story of a Missile* (New York: Harper & Brothers, 1960), 152–153.

[30] J. D. Hunley, ed., *The Birth of NASA*, 13.

Abe Silverstein, Chief of Research at Lewis Flight Propulsion Laboratory, 1950s. (C1963-63846)

For Dempsey, Centaur was the key to securing the company's place in the emerging aerospace industry. Though the Atlas missile became operational in 1959, it had severe limitations as a weapons system. The company anticipated its replacement in the country's arsenal with the Minuteman solid-propellant missile and the much larger Titan II. Atlas, fitted out with a new high-energy upper stage, later proved a commercial success for the company. However, development of Centaur into a dependable launch vehicle would be neither inexpensive nor easy. The difficulties of developing a liquid-hydrogen-fueled upper stage would prove a shocking revelation to the talented engineering team at General Dynamics.

The General Dynamics proposal, "A Satellite and Space Development Program," was ready two months after Sputnik. The group pitched the idea to Dempsey, who immediately packed Ehricke and William Patterson off on a sales mission to Washington, District of Columbia. They submitted Krafft Ehricke's first Centaur proposal to the Air Force in December 1957. The company asked for $15 million to begin work on the high-energy upper stage.[31]

Although the Air Force did not accept the first proposal, Ehricke called on Abe Silverstein at NACA Headquarters during the following June. Well known in the nation's aerospace community, Silverstein was among the nation's most knowledgeable experts in aircraft and rocket propulsion. Born in 1908 in Terre Haute, Indiana, Silverstein had graduated from Rose Polytechnic Institute in 1929 with a degree in mechanical engineering. After more than a decade in wind-tunnel design and operations at NACA's Langley Laboratory, he transferred to the new engine research laboratory in Cleveland during World War II. As Chief of Research for the Lewis Flight Propulsion Laboratory in Cleveland in the 1950s, he aggressively supported the work of the rocket section and put pressure on NACA Headquarters for greater research funding for the study of rocket propellants.

Silverstein shared Ehricke's enthusiasm for liquid hydrogen, but his hands were tied. Congress had not yet passed the Space Act setting up the National Aeronautics and Space Administration. He suggested that ARPA might be receptive. Ehricke presented the proposal for a liquid-hydrogen upper stage to Herbert York, who was chief scientist for ARPA at that time. Because of the well-known difficulties of developing a hydrogen pump, Ehricke had specified a pressure-fed system with four 7,500-pound-thrust oxygen/hydrogen engines to be designed by the Rocketdyne Division of North American Aviation.[32] At this meeting, he was surprised and delighted to learn of the Air Force's prior funding for the development of a liquid-hydrogen airplane for high-altitude reconnaissance.[33] At the ARPA meeting, Ehricke was told that the Air Force was so strongly committed to the development of liquid hydrogen that large liquefiers to convert hydrogen gas to liquid hydrogen had already been built in various parts of the country.

Early Liquid-Hydrogen Initiatives

Ehricke and the design team at General Dynamics were unaware of earlier liquid-hydrogen initiatives. Centaur brought together liquid-hydrogen research and development by academia, the military, and NACA. The development of a national capability in liquid hydrogen began in the 1920s with the pioneering research of William Giauque and his student Herrick L. Johnston at the University of California at Berkeley. Their research contributed to the discovery of deuterium, or heavy hydrogen, in 1931 by Nobel Prize winner Harold Urey. When Johnston took a position at The Ohio State University (OSU) in the late 1920s, he

---

[31] Patterson, 20.

[32] *Centaur Technical Handbook and Log*, General Dynamics, 12 March 1963, "Old RL10 Records," NASA Glenn Research Center (GRC) Records.

[33] Interview with Krafft Ehricke by John Sloop, 26 April 1974, NASA Historical Reference Collection.

continued this work in cryogenics. The hydrogen liquefier, or cryostat, he built at Ohio State provided a model for other cryostats built in the United States after World War II.[34] With new methods to convert hydrogen gas to liquid hydrogen, postwar rocket designers began to consider the tantalizing possibilities of liquid hydrogen. Not only could liquid hydrogen be stored in a smaller space than hydrogen gas, but keeping hydrogen in its liquid state until just prior to ignition significantly reduced the dangers of handling and storage.

Liquid-hydrogen research found early support from the Department of Defense. The Air Force sponsored a series of liquid-hydrogen rocket experiments by Johnston and his students in which they applied the principle of regenerative cooling by circulating liquid hydrogen in passages that surround the thrust chamber prior to injection and firing of the rocket.[35] They also discovered that ball bearings cooled by liquid hydrogen did not require lubrication—a phenomenon rediscovered by engineers at Pratt & Whitney during the development of the RL10 engine.

In the late 1940s, experiments at the Aerojet Engineering Corporation in Azusa, California, contributed to solving problems associated with liquid-hydrogen supply. With help from Johnston's team at OSU, Aerojet built a hydrogen liquefier.[36] When Aerojet lost Navy support, the work was continued by the Jet Propulsion Laboratory in Pasadena, where a liquid-hydrogen rocket was tested for the first time.[37] The Navy also contracted with Martin Marietta and North American Aviation to design a hydrogen/oxygen rocket called the High Altitude Test Vehicle (HATV). This design featured pressure-stabilized tanks with a common bulkhead separating the two propellants. None of these projects enjoyed high visibility or lavish funding.[38]

In the early 1950s, after President Harry Truman authorized the development of a hydrogen bomb, generous support for basic and applied cryogenics research suddenly became available to academic, government, and industry researchers.[39] In 1952, the U.S. National Bureau of Standards, under the auspices of the Atomic Energy Commission, set up a Cryogenics Engineering Laboratory in Boulder, Colorado, to produce liquid hydrogen in larger quantities. The discovery in 1953 of a catalyst to take free hydrogen, a mixture of one isotope of regular hydrogen and one of deuterium, and convert it to a para-hydrogen molecule before liquefaction greatly speeded up the conversion process.[40] The Boulder laboratory established a data center and began to publish

---

[34] John Sloop, *Liquid Hydrogen*, 13–15.

[35] John Sloop, *Liquid Hydrogen*, 23.

[36] See H. L. Coplen, "Large-Scale Production and Handling of Liquid Hydrogen," *Journal of the American Rocket Society* 22 (November–December 1952): 309–322, 338.

[37] John Sloop, *Liquid Hydrogen*, 54–55.

[38] John Sloop, *Liquid Hydrogen*, 44.

[39] See Richard Rhodes, *Dark Sun: The Making of the Hydrogen Bomb* (New York: Simon & Schuster, 1995), 487–489.

[40] See Joseph Green and Fuller C. Jones, "The Bugs That Live at -423°," *Analog Science Fiction/Science Fact*, 80 (January 1968): 14.

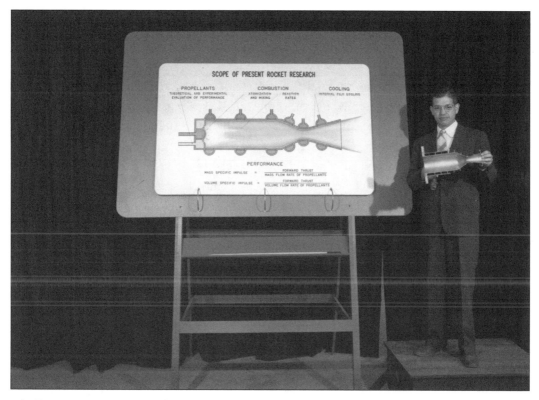

John Sloop presents scope of NACA rocket research, 1947. (NASA G19785)

the results of experimental research conducted there. However, production capacity from all hydrogen liquefiers in the United States was less than one-half ton per day as late as 1956.

An effort by NACA at the Lewis Flight Propulsion Laboratory in Cleveland was among the important early liquid-hydrogen initiatives that contributed to a national capability in high-energy fuels. Members of the Rocket section of the laboratory studied the performance of various liquid propellants and chamber and nozzle cooling.[41] After study of German technical papers, as well as reports produced by Aerojet and the Jet Propulsion Laboratory, they decided to narrow their investigation to one of the least studied areas, that of high-energy liquid propellants. An important 1948 theoretical study by Paul Ordin and Riley Miller concluded that liquid hydrogen held promise as a rocket fuel because of its high specific impulse.[42]

Because of the laboratory's lack of rocket test facilities and its inability to purchase liquid hydrogen in sufficiently large quantities, the group focused on then-exotic chemicals such as

---

[41] "Combustion Research Program, NACA Lewis Laboratory, Active as of September 1949," NASA Lewis Research Center Records, RG 255, Box 295, NARA, Chicago, IL. The Rocket Section was part of the Combustion Branch of the Fuels and Lubricants Division.

[42] R. O. Miller and P. M. Ordin, "Theoretical Performance of Rocket Propellants Containing Hydrogen, Nitrogen, and Oxygen," NACA RM E8A30 (1948), cited in John Sloop, *Liquid Hydrogen*, 80.

hydrazine with oxidizers like hydrogen peroxide, chlorine trifluoride, liquid oxygen, nitrogen tetroxide, and liquid fluorine.[43] Although a small research effort, this work on rocket fuels laid the basis for the Cleveland laboratory's rocket expertise.

By 1954, researchers had completed dozens of reports on various rocket fuels and components. They were experimenting with full-scale rocket engines and beginning to focus on hydrogen/oxygen as a promising propellant combination in a 5,000-pound rocket, but funding was inadequate to carry out extensive testing.

In late 1955, the Cleveland laboratory seized the opportunity to increase its expertise in the storage, handling, and firing of liquid hydrogen by contributing to an Air Force project to develop a more advanced spy plane than Lockheed's U-2, which flew at very high altitudes over the Soviet Union to avoid getting shot down. The low density of the atmosphere at high altitudes limited the performance of the U-2's engine, which used conventional kerosene-based fuel. Because hydrogen burned readily at extremely low pressures, it seemed a promising fuel for high-altitude reconnaissance aircraft. To investigate the possibilities of liquid-hydrogen fuel, Silverstein and NACA research engineer Eldon Hall produced a classified NACA technical report later published as "Liquid Hydrogen as a Jet Fuel for High-Altitude Aircraft."[44] The report recommended a pressure-stabilized structure for the design of an airplane fuel tank filled with liquid hydrogen. Whether the Silverstein-Hall report stimulated the Air Force's keen interest in liquid hydrogen or simply reinforced it is unclear. Silverstein won funding for a $1-million liquid-hydrogen test program at Lewis called Project Bee. The Air Force supplied a B-57 bomber and two Curtiss Wright J-35 engines for this research. The laboratory was given one year to determine whether the aircraft could be adapted to fly with liquid-hydrogen fuel.

Project Bee gave Lewis engineers experience with problems such as the insulation, instrumentation, and pumping of an engine run on liquid hydrogen. They determined that even when the fuel tank was insulated, heat leaks caused the temperature of liquid hydrogen to rise. When it reached the boiling point, the liquid fuel changed to hydrogen gas, causing the pressure in the tank to rise. Without proper venting, the fuel tank exploded. Too much venting, however, reduced the fuel available for completing the flight. They also discovered that helium gas prevented the insulation from freezing to the exterior wall of the tank.[45] During the first flight over Lake Erie in December 1956, the pilot switched from JP-4 fuel to liquid hydrogen twice, but the aircraft failed to maintain its

---

[43] John Sloop, *Liquid Hydrogen*, 75.

[44] NACA RM E55 C28a, 15 April 1955. See John Sloop, *Liquid Hydrogen*, 98–102.

[45] See "Experimental Study of Foam-Insulated Liquefied Gas Tanks," by Thaine W. Reynolds and Solomon Weiss, NACA RM-E56K08A.

speed. During a subsequent flight, the engine operated for 20 minutes on liquid hydrogen at an altitude of 49,500 feet and a speed of Mach 0.72.[46] By the mid-1950s, the Lewis laboratory had become one of the acknowledged leaders in cryogenic fuels research, but it lacked adequate facilities for rocket testing.[47]

Centaur and Project Suntan

Project Suntan, the code name for a top-secret, high-altitude spy plane, directly contributed to the development of Centaur.[48] During Project Suntan, the Air Force funded the construction of hydrogen-liquefaction facilities in Trenton, New Jersey (later canceled); Painesville, Ohio; and Bakersfield, California. Two additional liquefiers were built adjacent to a new Pratt & Whitney test center in West Palm Beach, Florida. Together, these plants increased the capacity of the United States to produce liquid hydrogen from 500 pounds a day in 1956 to about 68,000 pounds a day by 1959. This liquid-hydrogen infrastructure contributed to the national capability to develop a liquid-hydrogen rocket.

Another important connection between Suntan and Centaur was its Air Force management. John Seaberg, later one of Centaur's champions, headed Project Suntan. A veteran of the U.S. Army Air Corps, Seaberg had worked for Chance Vought Aircraft for five years as an aerodynamicist until the start of the Korean War, when he was recalled to active duty at Wright-Patterson Air Force Base in Dayton, Ohio. Seaberg conceived and directed the U-2 program. A strong proponent of liquid hydrogen, Seaberg obtained funding for an advanced spy plane designed by Clarence (Kelly) Johnson of Lockheed in 1956. Pratt & Whitney Aircraft Engine Company received a contract to design Suntan's liquid-hydrogen turbojet engine. Pratt & Whitney's design team, headed by Richard Coar, completed a prototype for the innovative 304 turbojet engine in just sixteen months.

In conventional turbojet engines, a small fraction of the propellants is burned in a separate combustion chamber to begin the combustion process. This requires a separate propellant control and ignition system, which adds complexity and weight. The Pratt & Whitney engine had an "expander cycle" in which liquid hydrogen flowed through tubes into the combustion chamber walls on the way to the turbine. In the process, the cryogenic propellant, now gaseous,

---

[46] Interview with Paul Ordin by John Sloop, 30 May 1974, NASA Historical Reference Collection. See also John L. Sloop, "NACA High Energy Rocket Propellant Research in the Fifties," AIAA 8th Annual Meeting, 28 October 1971, NASA Historical Reference Collection.

[47] Richard Canright, "Rocket Propellants: Views of an Airframe Manufacturer," 31 July 1957, transmitted to members of the NACA Subcommittee on Rocket Engines with a cover letter by Benson E. Gammon, 21 August 1957, quoted in Sloop, *Liquid Hydrogen*, 85.

[48] See John Sloop, *Liquid Hydrogen*, on project Suntan, 141–166, and "General Discussion by Col. N. C. Appold, Lt. Col. J. D. Seaberg, Maj. A. J. Gardner, Capt. J. R. Brill," transcribed February 1959, 15, John Sloop Papers, NASA Historical Reference Collection.

picked up heat and expanded, providing energy to turn the turbine and turbopumps; at the same time, it cooled the thrust chamber walls.

Despite the elegance of the Pratt & Whitney design for the engine, the design team at Lockheed began to doubt that the proposed altitude and range for the new airplane could be achieved. Kelly Johnson convinced Donald Quarles, Deputy Secretary of Defense, to cancel the project. Nevertheless, Lieutenant Colonel Seaberg continued to champion liquid hydrogen. In June 1958, he convinced ARPA officials to fund a project first known simply as the high-energy upper stage. Thus, when Ehricke coincidentally presented his plans for a liquid-hydrogen upper stage for Atlas, the proposal won immediate support. John Sloop commented, "Two days after the ARPA order, Pratt & Whitney conducted the tenth and final series of tests with the hydrogen fueled 304 turbojet engine. Suntan became a thing of the past and Centaur, a hydrogen-oxygen rocket stage on top of Atlas, rose as its replacement. All the plant equipment, and technology of Suntan could now be brought to bear in assuring that Centaur would succeed."[49]

Project management was assigned to a special projects office at Wright Field under Colonel Norman Appold until Abe Silverstein, in charge of the Office of Space Flight Programs for the newly formed space agency, insisted that NASA take over the program. However, the Air Force officers assigned to project continued to manage it under NASA's direction. This arrangement later became a source of friction between the Air Force and Marshall Space Flight Center.

Ehricke learned of Pratt & Whitney's liquid-hydrogen engine at his first meeting with ARPA officials in the summer of 1958. The innovative expander cycle of the engine proved a revelation to Ehricke and his colleagues at General Dynamics. Kenneth Newton, Director of Launch Vehicle Programs, recalled the excitement that the first set of Pratt & Whitney drawings generated at the company. "We were so enthralled with their simplicity," he said.[50] The expander cycle (also referred to as a "bootstrap cycle") of the Suntan engine would serve as the basis for the design of the Centaur RL10 engine (originally designated LR115). It was one of the keys to the striking efficiency and reliability of the RL10. Dick Mulready, second in command of the RL10 project, wrote:

> It is hard to over-emphasize the significance of the cycle choice for the RL10. The singular physical properties of pure hydrogen, which the cycle exploits, allow very large margins of safety in design. The simple cycle completely eliminates many complex ancillary systems. This fact and the large design margins that are possible at low temperature lead to unmatched reliability. Starting is accomplished by opening the propellant

---

[49] John Sloop, *Liquid Hydrogen*, 196. See also Roger Bilstein, *Stages to Saturn*, 135. The first public announcement of the project was made by General Bernard Schriever in May 1959. See R. B. Scott, *Technology and Uses of Liquid Hydrogen* (New York: The Macmillan Company, 1964), 3.

[50] Interview with the Aerospace Division of Convair/General Dynamics by John Sloop, 29 April 1974, NASA Historical Reference Collection.

supply valves and admitting hydrogen into the thrust chamber cooling tubes. The sensible heat residual in the tubes, even after many hours of coasting in space, is enough to spin up the turbine.[51]

The unusual design of the RL10 engine, based on the unique properties of liquid hydrogen, prompted Ehricke to propose a bold new design for Centaur to capitalize on the low density and extreme cold of liquid hydrogen. Ehricke had first envisioned a more complicated design for Centaur consisting of two large tanks—one for liquid hydrogen and the other for liquid oxygen. Each had a smaller tank within the larger one. The inner tanks—intended to supply fuel to restart the engines in space after a coast period—added weight and complexity.[52]

Ehricke simplified the design for Centaur. By adopting Pratt & Whitney's hydrogen pump, two of the four proposed engines could be eliminated. Rocketdyne, the country's preeminent designer of rocket engines, was outraged that the Air Force had awarded a sole-source contract to Pratt & Whitney, a company without prior experience in the design of a rocket engine.[53] The decision was based on Pratt & Whitney's previous experience with liquid hydrogen and the propellant's ready availability at the Florida test site.

Next, Ehricke eliminated the unnecessary complexity of the prior design. The new Centaur was remarkably like the thin-walled, pressure-stabilized Atlas in structure. Centaur's two propellants were separated by a double-walled bulkhead, consisting of two thin metal skins separated by a quarter-inch layer of fiberglass insulation. The internal pressure of the two liquids kept the outer wall of the tank rigid and pushed against the bulkhead to keep it from collapsing. Prior to tanking with its cryogenic propellants, the bulkhead was backfilled with dry gaseous nitrogen. When loaded with liquid oxygen on one side of the bulkhead and liquid hydrogen on the other, the liquid hydrogen froze out the nitrogen to produce a near-perfect vacuum within the hollow bulkhead. This thermal barrier greatly reduced heat transfer between the liquid oxygen (-299°F) and the much colder liquid hydrogen (-423°F). Pratt & Whitney's expander cycle and Ehricke's double-walled bulkhead design were both logical extensions of the special properties of liquid hydrogen. Ehricke said:

> Using this design, we achieve a very high mass ratio for a vehicle of this size with low-density propellants. The membrane bulkhead design uses cryogenic vacuum formation as the logical utilization of an existing extreme temperature environment; just like Pratt

---

[51] Dick Mulready, *Advanced Engine Development at Pratt & Whitney*, 60. See also Joel E. Tucker, "The History of the RL10 Upper-Stage Rocket Engine," 126.

[52] Interview with the Aerospace Division of Convair/General Dynamics by John Sloop, 29 April 1974, NASA Historical Reference Collection. Ehricke also mentioned the earlier design in his 1962 congressional testimony. Neither the original proposal nor a blueprint of the design is available to confirm inclusion of a pressure-stabilized structure.

[53] Robert S. Kraemer, *Beyond the Moon: A Golden Age of Planetary Exploration 1971–1978* (Washington: Smithsonian Institution Press, 2000), 25.

& Whitney's bootstrap method of engine start [i.e., use of residual heat in the thrust chamber cooling tubes to spin the turbine] was an ingenious and logical utilization of the special physical characteristics of liquid hydrogen.[54]

The double-walled bulkhead, created late one night with the help of the company's thermodynamicist, William Mitchell, was arguably the most fundamental part of the design. "We had that integral bulkhead before we even knew how many engines we had on the thing or how long the thing was going to be. That was the kind of invention that sprung the whole thing," Deane Davis, Centaur project engineer, recalled.[55] The bulkhead restricted heat transfer between the relatively warmer liquid oxygen and the liquid hydrogen. The double-walled bulkhead would not only be the most innovative feature of Centaur, but also the most technically demanding. Engineers would later discover that liquid hydrogen leaked through minute pores in the welds of the bulkhead.[56]

Another innovative feature involved putting the thrust structure, which distributes the concentrated thrust load, inside the liquid-oxygen tank. This saved weight by shortening the vehicle and reducing bending loads. The exterior of the liquid-hydrogen tank was protected by four quarter-section fiberglass insulation panels. The insulation shielded the liquid hydrogen from aerodynamic heating, minimizing fuel boil-off. The panels remained bolted together during the early portion of flight. As soon as the aerodynamic heating rate was no longer critical, shaped explosive charges were fired to separate them from the vehicle. The design also included explosive bolts to separate the two halves of the nose fairing. Small nitrogen gas thrusters would propel the two sections of the fairing away from the vehicle.[57]

To make sure that nothing interfered with the General Dynamics focus on Atlas, ARPA funded Centaur as an experimental program without a mission assigned to it. Ehricke's new proposal, accepted in August 1958, had modest goals. Considered a bridge between Atlas-Agena and much larger future boosters, Centaur was intended to prove the feasibility of a liquid-hydrogen/liquid-oxygen rocket. The guidance system, test facilities, and launch complex were not included in the bare-bones contract. Development was proposed to take twenty-five months, with the first flight to occur in January 1961. Cost was projected at $36 million for six vehicles.

To keep down costs and prevent interference with the Atlas program, ARPA insisted that General Dynamics use Atlas hardware wherever possible.[58] A member of the design team

---

[54] U.S. House of Representatives Subcommittee on Space Sciences of the Committee on Science and Astronautics, "The Centaur Program," 87th Congress, 15 and 18 May 1962, 67.

[55] Interview with the Aerospace Division of Convair/General Dynamics by John Sloop, 29 April 1974.

[56] Richard E. Martin, "The Atlas and Centaur 'Steel Balloon' Tanks: A Legacy of Karel Bossart," reprint by General Dynamics Corp., 40th International Astronautical Congress paper, IAA-89-738, 1989, 11.

[57] Description from *Centaur Capability Handbook*, GD/A-BTD64-119-1, 1 October 1964, Kennedy Space Center ELV Resource Library.

[58] Centaur Program, 1962 Hearings, 66.

Centaur is raised into the "J" Tower for testing at Point Loma, early 1960s. (Courtesy of Lockheed Martin)

recalled, "One of the ground rules to us was to use Atlas tooling wherever possible, and so really the Centaur was, and still is, bits and pieces of Atlas put together."[59] These design constraints sometimes interfered with achieving an optimal design.

Although a NASA project, Centaur continued to be managed by the Air Force. Ehricke's monthly reports indicated the growing complexity of the project. Program costs increased to $42 million with the addition of a guidance system and construction of additional test facilities. By June 1959, the company had awarded a subcontract for the inertial guidance system to Minneapolis-Honeywell. In July, General Dynamics signed a contract for construction of a launch complex, designated Pad 36, at Cape Canaveral. In August, it completed construction of a flow test facility at Point Loma, located on a peninsula on the Pacific coast north of San Diego. A static test facility in Sycamore Canyon was activated in September shortly before the first test. The Centaur team then initiated liquid-hydrogen tanking tests at Point Loma.

Glennan took steps to prevent any disruption of Centaur development when it was transferred from ARPA to NASA in July 1959. He urged that the project office be located in Washington, but he was overruled by officials of the Department of Defense who insisted that Centaur management be coordinated by the Air Force from its Ballistic Missile Division at Inglewood, near Los Angeles, California.

Because the United States had no launch vehicle powerful enough to lift the heavy satellites needed for a global military communications network, Centaur became one of the top priorities of the military. Herbert York urged Glennan to expedite the program because it was needed for launching the Advent series of communications satellites assigned to Centaur in December 1959.[60]

A secret five-year $174-million research and development program, Advent included ten planned launches of communications satellites for the Army, Air Force, and Navy. ARPA was counting on the 1,200-pound satellites to provide instant point-to-point transmissions to receiving stations within the line-of-sight beam of the satellite. The mission required restarting the Centaur engines in space. After insertion into a nearly 1-hour parking orbit, Centaur would position the satellite for an elliptical 5-hour coast orbit. The final restart would then place the spacecraft in a 24-hour circular orbit. Since Centaur was not yet ready, the first three satellites were supposed to be launched with the far less powerful Agena B as the upper stage for Atlas. The Air Force planned to launch the remaining seven Advent

---

[59] Interview with the Aerospace Division of Convair/General Dynamics by John Sloop, 29 April 1974, NASA Historical Reference Collection.

[60] Herbert York to T. Keith Glennan, 19 September 1959, Centaur files, NASA Historical Reference Collection. On Advent, see David N. Spires and Rick W. Sturdevant, "From Advent to Milstar: The U.S. Air Force and the Challenges of Military Satellite Communications," in *Beyond the Ionosphere: Fifty years of Satellite Communication*, ed. Andrew J. Butrica (Washington, DC: NASA SP-4217, 1997), 65.

satellites in 1963 on Atlas-Centaur. Centaur's development problems made meeting the demands of this schedule impossible.[61]

By December 1959, General Dynamics had completed a metal prototype and moved it to the final assembly area, where a mockup of the electrical and pumping systems could be added. At the same time, the company began a flight program at Edwards Air Force Base to investigate the behavior of liquid hydrogen under zero-gravity conditions using an Air Force KC-135. Ehricke optimistically predicted that the first Centaur flight article would be delivered in January 1961.[62]

Centaur and Saturn

In addition to the assignment of Advent to Centaur, another key decision of the early space program took place in December 1959. Six months before the official transfer of the von Braun team to NASA, Abe Silverstein chaired a committee to evaluate the Saturn vehicle and settle the question of propellants for Saturn's upper stages.[63] Silverstein occupied one of NASA's most important posts during NASA's formative years. His Office of Space Flight Programs was responsible for both human and robotic exploration of space. This office set the agenda for planetary exploration and completed preliminary planning for a lunar landing well before it became a goal of the Kennedy administration.

Silverstein had been determined to prove the competence of former NACA engineers to run the new agency when he accepted the job at NACA Headquarters in 1957.[64] He personally recruited many of the NASA Headquarters staff, bringing a large contingent from Lewis Research Center in Cleveland. Because the inadequacy of the nation's launch vehicles was a source of concern at all levels of government, Silverstein championed Centaur and called it "the kind of thing upon which our whole future technology, I think, rests—that is, the development of an early capability with these high-energy propellants."[65] Based on Silverstein's advocacy,

---

[61] "Military Space Projects: Report of Progress for June–July 1960," Office of the Director of Defense Research and Engineering," Office of the Staff Secretary: Records of Paul T. Carroll et al., 1952–1961, Department of Defense Subseries, Box 9, "Missiles and Satellites, Military Space Projects [June–August 1960]," Dwight D. Eisenhower Library. The authors thank Dwayne Day for providing a copy of this document.

[62] Centaur Progress Reports, 31 December 1958 through 31 July 1960, San Diego Aerospace Museum Archives.

[63] See John Sloop, *Liquid Hydrogen*, 230–243. Other members included Abraham Hyatt, NASA; George P. Sutton, ARPA; T. C. Muse, Office of the Director of Defense Research and Engineering (ODDR&E); Norman C. Appold, U.S. Air Force; and Wernher von Braun, Army Ballistic Missile Agency (ABMA). Eldon Hall, Silverstein's close colleague at NASA, served as secretary. See "Report to the Administrator on Saturn Development Plan by Saturn Vehicle Team," 15 December 1959, Silverstein file, NASA Historical Reference Collection. The text of the report is partially reproduced in John M. Logsdon et al., *Exploring the Unknown, Volume IV: Accessing Space*, document 1-29, 116–119.

[64] Homer E. Newell, *Beyond the Atmosphere: Early Years of Space Science* (NASA SP-4211, 1980), 100.

[65] U.S. House of Representatives, *1960 NASA Authorization*, Hearings Before the Committee on Science and Astronautics, 86th Congress, H.R. 6512, April and May 1959, 391.

Congress authorized additional funding and increased the contract with General Dynamics from six to ten vehicles—"this being considered an absolute minimum to prove out a vehicle design upon which so much of the national space program was beginning to depend."[66]

Because of Silverstein's strong faith in liquid hydrogen, he insisted that Saturn include at least one liquid-hydrogen upper stage. The team, headed by Wernher von Braun, was adamantly opposed to liquid hydrogen. At Kummersdorf in the 1930s, von Braun had witnessed the demonstration of a small liquid-hydrogen engine designed by Walter Thiel and noted that "the greatest impression he retained was of the numerous line leaks and difficulties of handling liquid hydrogen."[67]

At a series of key meetings of the Silverstein committee, von Braun urged the use of conventional kerosene-based fuels. In Silverstein's view, "The von Braun team was apparently willing to take on the difficulties of the 1.5 million-pound thrust-booster stage rather than the hazards which they contemplated in the use of hydrogen as fuel."[68] Silverstein used data prepared by NACA colleague Eldon Hall to beat down opposition to liquid hydrogen. For a week, the NASA team and the von Braun group debated, until von Braun, to the astonishment of the other members of his team, capitulated. The Saturn Launch Vehicle task team was unanimous in its endorsement of liquid hydrogen/liquid oxygen in all Saturn upper stages. Their report concluded, "Current success in the Centaur engine program substantiates the choice of hydrogen and oxygen for the high-energy propellants."[69] To Oswald Lange, project director for Saturn at Marshall, the recommendations of the Silverstein committee for liquid-hydrogen upper stages represented a "major milestone in the Saturn program."[70] The Saturn C-1 would use conventional fuel in its first stage but have liquid-hydrogen upper stages. The second stage (designated S-IV) was to use four RL10 engines uprated to 20,000 pounds thrust each. The proposed C-1's third stage would be the Centaur itself with two RL10 engines, each with 15,000 pounds of thrust. Silverstein said in a 1977 interview that he did not know exactly what led von Braun to accept liquid hydrogen, but in hindsight it had proved to be the correct action. "I believe that the decision to go with hydrogen-oxygen in the upper stages of the Saturn V was

---

[66] Centaur Program, 1962 hearings, 7. According to *NASA Pocket Statistics* for December 1961, research and development (R&D) funding for Centaur increased from $4 million in FY 1959 to $36.64 million in FY 1960 and to $62.58 million in FY 1961.

[67] John Sloop, *Liquid Hydrogen*, 236.

[68] Abe Silverstein, "How It All Began," speech at Kennedy Space Center Writers Conference, Cocoa Beach, Florida, 1–3 September 1977, quoted by Virginia Dawson, *Engines and Innovation*, 167.

[69] "Report on Saturn Development Plan B Saturn Vehicle Team," 15 December 1959, Silverstein file, NASA Historical Reference Collection.

[70] Oswald H. Lange, "Development of the Saturn Space Carrier Vehicle," in Ernst Stuhlinger et al., *Astronautical Engineering and Science from Peenemünde to Planetary Space* (New York: McGraw-Hill Book Company, Inc., 1963), 5. See also the letter from Abraham Hyatt to Thomas O. Paine, 24 November 1969, Silverstein family papers. NASA Glenn History Office.

the significant technical decision that enabled the United States to achieve the first manned lunar landing," he said. "The Russian effort to accomplish this mission without high-energy upper stages was doomed to failure."[71]

The decision of the Silverstein committee directly influenced the fortunes of Centaur. Centaur moved into NASA's mainstream as a stepping-stone to a piloted lunar landing. However, the von Braun team, although convinced of the superiority of liquid-hydrogen upper stages, was never fully committed to Centaur. Referring to the new designation of Centaur as the final stage (or S-V stage) for the Saturn C-1, Krafft Ehricke joked that on top of Saturn's "horrible tower," at least there would be one "decent vehicle." He referred to the Saturn C-1 as the "Centaur-tipped Saturn," to which von Braun retorted, "Centaur-tipped Saturn, Hell!"[72]

Von Braun favored large, heavy structures to ensure reliability. His conservative design philosophy, which Ehricke derided as comparable to the Brooklyn Bridge in its conservatism, "mitigated against the use of liquid hydrogen which, more than conventional fuels, depended upon very light structures to help offset the handicap of low density."[73] Though von Braun agreed to Saturn stages powered by liquid hydrogen, he apparently did not approve of Centaur's balloon structure. His opposition may have influenced the NASA decision to award the contract for the S-IV second stage to the Douglas Aircraft Company, ostensibly to provide "some semblance of competition" among aerospace companies.[74] The S-IV second stage, with its cluster of six RL10 engines, produced an additional 90,000 pounds of thrust to the 1,296,000 pounds of thrust generated by the first stage. The choice of Douglas was controversial enough to merit an investigation by the GAO, since Douglas had no prior experience with liquid hydrogen.[75] Douglas avoided the balloon structure by adopting a honeycomb design similar to that of the Douglas Thor missile.

Caught Between DOD and NASA

From July 1960 to early 1962, the Air Force continued to manage both the Centaur and the Agena programs, reporting to Marshall Space Flight Center. D. L. Forsythe, the Agena program chief,

---

[71] See note 68.

[72] Interview with Krafft Ehricke by John Sloop, 26 April 1974, NASA Historical Reference Collection.

[73] John Sloop, *Liquid Hydrogen*, 208. See also G. R. Richards and Joel W. Powell, "The Centaur Vehicle," *Journal of the British Interplanetary Society* 42 (1989): 99–120.

[74] J. D. Hunley, *The Birth of NASA*, 129. See also Ray A. Williamson, "The Biggest of Them All, Reconsidering the Saturn V," in *To Reach the High Frontier*, 304.

[75] Ray Williamson, "Access to Space: Steps to the Saturn V," in John M. Logsdon et al., *Exploring the Unknown, Volume IV, Accessing Space* (Washington, DC: NASA SP-4407): 1–31.

lamented the lack of Marshall attentiveness to both Agena and Centaur, which he said received "only a few crumbs which have fallen from the banquet table of thought and effort at MSFC."[76]

NASA had acquired the Agena rocket, manufactured by the Lockheed Missiles and Space Company, in early 1960. The engine, manufactured by Bell Aerospace Company, burned unsymmetrical dimethylhydrazine with inhibited red fuming nitric acid as the oxidizer. As a second stage for Atlas, Agena could launch only about 500 to 750 pounds into an Earth-escape trajectory. Although the Air Force needed Agena to launch the early Advent satellites, the Agena program proved an unwanted distraction to Marshall engineers whose energies were consumed by Saturn development.

Although Marshall had little interest in Centaur, the RL10 engine figured prominently in its plans because von Braun was counting on an uprated version of the engine for the Saturn S-IV liquid-hydrogen upper stage. In February 1960, he began to lobby to take over RL10 engine management from the Air Force. This made sense to A. O. Tischler, NASA's Chief of Liquid Fuel Rocket Engines, because the Air Force seemed incapable of keeping development costs under control. The RL10 engine already had a project overrun of about $10 million. The danger in allowing Marshall to take over the RL10 engine contract was that the engine for Centaur might be neglected in the interest of the hydrogen engine for the Saturn stage. He urged that Headquarters management be strengthened in order to minimize project favoritism.[77]

Major General Don Richard Ostrander, Director of NASA's launch vehicle programs between 1959 and 1961, had similar concerns. He told von Braun that he was concerned that the high cost of RL10 engine development would draw funds away from other parts of the Saturn program because of the Pratt & Whitney reputation for high costs. But he admitted that Pratt & Whitney produced engines of unusually high quality.[78] Immediately after Marshall won its bid to take over RL10 development, it contracted with Pratt & Whitney for a more powerful version of the RL10 called the LR-119. The Douglas S-IV stage specified a four-engine cluster producing 70,000 pounds of thrust. Later, the configuration was changed to a six-engine cluster producing 90,000 pounds of thrust.[79]

Centaur's prominence within NASA increased when it was assigned two important planetary and lunar exploration programs, Mariner in 1959 and Surveyor in 1961. Engineers at the Jet Propulsion Laboratory in Pasadena, California, who managed both Surveyor and Mariner, were skeptical of Centaur's vaunted advantages over more conventional propulsion

---

[76] Edward Clinton Ezell and Linda Neuman Ezell, *On Mars: Exploration of the Red Planet* (NASA SP-4212, 1984), 47.

[77] A. O. Tischler to Abe Hyatt, 5 February 1960, John Sloop papers, NASA Historical Reference Collection.

[78] Don R. Ostrander to Wernher von Braun, 20 June 1960, John Sloop papers, NASA Historical Reference Collection.

[79] See Joel E. Tucker, "The History of the RL 10 Upper-Stage rocket Engine, 1956–1980," published in *History of Liquid Rocket Engine Development in the United States, 1955–1980*, ed. Stephen E. Doyle, AAS History Series, vol. 13 (San Diego: AAS, 1992), 133. See also Mulready, 78; Bilstein, 131–140, 188–190.

systems. Their opposition to Centaur would grow as Centaur's development problems became increasingly apparent.

NASA set up a Centaur Project Technical Team under Colonel Don Heaton to facilitate technical support of NASA missions, although Centaur still remained under Air Force management because of its key role in the Advent Program for the Department of Defense. When Milton Rosen took over as NASA's Chief of Rocket Vehicle Development, pressure on Ehricke increased. Rosen, who had served as project engineer for the Vanguard program, was not a fan of Centaur. He demanded that General Dynamics develop Vega as an upper stage for the Atlas. The Vega program, later canceled, siphoned off personnel and funds needed for the more technically challenging Centaur.[80]

General Dynamics began to build a flight-qualified Centaur, with the first launch now projected for June 1961. Construction was initiated on the Centaur launch complex at the Atlantic Missile Range at Cape Canaveral. A test program at the Tullahoma Vacuum Test Facility at Arnold Engineering Development Center, additional nose-fairing development, and a zero-gravity flight-test program with Aerobee rockets at Wallops Flight Research Station all added to the increasing cost and complexity of the program.

NASA relied on Lewis Research Center's expertise in problems associated with the behavior of hydrogen in zero-gravity. The urgency of the program made it necessary to place fifty-four people at Lewis on a 48-hour week for a period of six months. The Center pressed the Propulsion Systems Laboratory into service for tank insulation and pressurization tests. Testing brought the disquieting knowledge that Centaur's intermediate bulkhead would collapse without the maintenance of a 3-pound pressure differential.

The Advent requirement that Centaur's engines be restarted twice after two coast periods posed questions for which there were not yet answers. During both periods, the longitudinal axis of the vehicle would be pointed at the Sun with the oxygen tank exposed to the Sun's heat. Also, depending on the time of year, Centaur might encounter other problems, such as radiation from the Earth. How would the propellants behave in the extreme environment of space, particularly when Centaur tanks were only partially filled? Would liquid hydrogen continue to completely wet the tank walls under near-zero-gravity conditions? Would the pressure buildup cause excessive venting, resulting in the loss of sufficient fuel for the second burn? Everyone involved in the space program—both NASA engineers and the people working under Krafft Ehricke at General Dynamics—were attempting to pierce the veil of ignorance surrounding the problems to be encountered in the low gravity of outer space.

One of the obstacles to piercing that veil was the design constraint of using Atlas tooling rather than designing Centaur from scratch. Fred Merino, a scientist at General Dynamics with a background in thermodynamics, recalls going to the design group headed by Daniel Heald.

---

[80] Interview with Donald Lesney (former General Dynamics employee) by Virginia Dawson, 5 June 2000.

Merino asked the group to design a special valve system for venting hydrogen gas during the two coast periods. They refused until Merino insisted that without proper venting the rocket would fail. He recalled:

> The designer responsible for some of this hardware said, "You can't do it because the ground rules are we use off-the-shelf hardware and they said that means we are going to take this Atlas valve, scale it down a little bit, mount the same valve on the liquid hydrogen tank as the liquid oxygen tank, and we're going to command the valve to open when you have to vent. That's all we're going to do." And I said, "You will never fly a mission that way. It's going to fail." The guy said, "Those are my marching orders." So I had to write a report explaining why upper management had to alter their design guidelines. They agreed when they heard why you can't do this.[81]

Denny Huber, an engineer hired by General Dynamics in 1960, recalled that the same bare-bones approach to Centaur tooling meant they had to use the Atlas liquid-oxygen propellant loading system. Although the new hydrogen system they had designed worked beautifully, the old oxygen system was inadequate. They had to wait until a second pad was built to redesign the system. Then Pad 36A was retrofitted with the new technology.[82]

Another question centered on how to insulate the Centaur liquid-hydrogen tank. Insulation was necessary to keep the liquid hydrogen from vaporizing on the ground and during launch. The insulation was applied in four segments running the length of the tank and was held against the tank by metal bands. At about ten times the speed of sound, explosive bolts on the bands were supposed to fire, allowing the insulation to fall away from the vehicle.[83] Lewis engineers were skeptical that the proposed insulation would work. A March 1961 memo by Lewis engineer Vern Gray stated, "The troubles with the insulation are delaying the test program, are potentially hazardous, and illustrate Convair's apparent indifference to the peculiar problems of hydrogen tank insulation." The memo described how hard it was to "cajole" test data out of the company. Gray observed on a visit to San Diego that the application of the insulation was uneven, leaving lumps. During tests, the insulation split open when the extreme cold of the tank wall came into contact with the air. The loss of insulation increased the rate of hydrogen boil-off, raising pressure in the tank. There was danger that the tank might rupture, spilling volatile liquid hydrogen. The effort to fix the problem with adhesive tape appalled

[81] Interview with Frederick Merino (former General Dynamics employee) by Virginia Dawson, 5 June 2000.

[82] E-mail communication from Denny Huber to Ed Bock, 30 March 2002.

[83] Memo from Irving Pinkel to Colonel D. H. Heaton, 25 January 1960, regarding "Meeting on Centaur insulation panel flutter, 22 January 1960," RG 255, NARA Box 254.

Gray. He commented, "Apparently, Convair has not grasped the importance of sealing off atmospheric air from a hydrogen tank nor the knowledge of how to do it."[84]

Representatives from the National Bureau of Standards Laboratory in Boulder, Colorado; Linde Air Products Company (with extensive knowledge of the handling of liquid oxygen); and NASA all voiced their frustration at "repeated refusals to accept suggestions for improvement." They suggested changing the type of insulation or applying an external seal coating, and purging beneath the shell with helium gas in order to prevent air from contaminating the insulation. The memo from Gray concluded, "Nothing being done now gives any real basis for optimism."[85]

As the costs of the program mounted and the launch date slipped, Centaur's troubles, both technical and managerial, became increasingly serious. Ehricke recalled his frustration with Pratt & Whitney and the Jet Propulsion Laboratory (JPL). Both organizations blamed General Dynamics for the delays in the Surveyor Program when, in fact, each had serious development problems of its own.[86] Up to this point, Marshall had left day-to-day management of the Centaur to the Air Force. But as Centaur problems multiplied, Headquarters demanded greater accountability. Advent, an important Department of Defense program, and two high-profile NASA missions, Mariner and Surveyor, depended on Centaur. Even more important, Centaur was considered essential to proving the feasibility of liquid-hydrogen upper stages for the Saturn C-1. Headquarters demanded that the von Braun team assert greater control over one of NASA's most innovative programs.

---

[84] Memo from Vernon H. Gray to Acting Director, 17 March 1961, RG 255, NARA 254.

[85] Ibid.

[86] Interview with Krafft Ehricke by John Sloop, 26 April 1974, NASA Historical Reference Collection.

Chapter 2

Marshall's Unruly Wards

"To the Huntsville people the methods and philosophies of the Atlas and Centaur were as mysterious as the dark side of the Moon."

–Deane Davis, General Dynamics

The Centaur had serious technical and managerial problems that proved an unwanted distraction from the development of the Saturn C-1 at Marshall Space Flight Center. Administrator Keith Glennan aptly captured Huntsville's attitude when he remarked, "Saturn was a dream; Centaur was a job."[1] As the Centaur schedule slipped, tensions between Marshall and NASA Headquarters surfaced. In January 1962, Headquarters insisted that von Braun become more involved in the troubled program. The explosion of the first Centaur in May 1962, a scathing indictment of the program by the House Subcommittee on Space Sciences, and bad press led to the transfer of Centaur from Huntsville to Cleveland in the fall of 1962. Implicit in the Centaur failure hearings was the tension between different philosophies of design and management of large-scale technology projects.

Huntsville's Indifference

As early as March 1961, Headquarters expressed increasing impatience with von Braun's indifference to the fate of Centaur. Don Ostrander, Director of NASA's launch vehicle programs, began to pressure von Braun to exercise greater oversight of the program. He presented him with a long confidential memo that almost certainly raised the hackles of the Huntsville team. It included a management checklist. Ostrander called Centaur "the most urgent task before us at this time." It was needed not only for Surveyor and Mariner, but also for "its hardware contribution to Saturn." Ostrander demanded that Centaur be "supported fully by the best talent available at MSFC." He instructed von Braun not only to submit informal monthly progress reports to Headquarters, but also to deliver them in person. He wanted regular meetings set up. These were to be attended by the senior NASA oversight personnel at Pratt & Whitney and General Dynamics: "Streamlined, effective management of

---

[1] Oran Nicks, *Far Travelers* (Washington, DC: NASA SP-480, 1985), 90.

the project must be considered essential," he wrote, "and any organization changes within NASA or within contractors' plants deemed necessary to accomplish this must be immediately implemented."[2] It appears that von Braun had no intention of assigning his best people to monitor General Dynamics.

Ostrander also thought that Pratt & Whitney needed better supervision. He wanted a test program set up at Lewis Research Center to help solve engine ignition problems. In short, he thought that Marshall must provide more active management of all aspects of the program, including a full-time project manager. To facilitate better management, the Air Force transferred all Centaur contracts (with the exception of those with Pratt & Whitney) to Marshall in June 1961. Von Braun named Hans Hueter Director of the Light and Medium Vehicle Office. Frances Evans took over as Centaur program manager.

These measures did not improve the performance of General Dynamics. In September 1961, the Centaur launch schedule continued to slip, while program costs rose to $100 million. In a confidential memo, Abe Silverstein informed Associate Administrator Robert Seamans that Centaur had become "an emergency of major proportions."[3] Silverstein emphasized that NASA had "a tremendous investment to protect." He prophetically warned, "More serious problems, such as explosion on the pad or a fundamental vehicle or propulsion problem, could virtually wipe out the robotic lunar and planetary program until 1965, except for the makeshift effort with Atlas-Agenas." Silverstein thought one of the main problems was that the Department of Defense Advent satellite seemed to be taking priority over NASA missions. Because Advent weight requirements were increasing, while the lunar and planetary requirements were holding steady, NASA was "saddled with an improvement program which we don't really need, one which will probably lower the reliability of the early NASA vehicles." He recommended that all changes to Surveyor be halted as soon as an injection weight of 2,500 pounds could be guaranteed. He also thought the RL10 engine allocations to Saturn should be reviewed to see whether they were interfering with Centaur. In November 1961, Homer Newell became head of a new Office of Space Sciences and Applications with responsibility for both Centaur and Surveyor. Newell, however, had limited time to coordinate the two extremely demanding programs.

Oran Nicks, Director of NASA's Lunar and Planetary Programs, was fully aware of the magnitude of Centaur's problems. He was dubious that Centaur would be ready in time to fly the Venus probe planned for August 1962. By the spring of 1961, he began to consider Atlas-Agena as an alternative. After discussing the possibility with people at JPL, he returned to Washington determined to find out whether the development of Centaur was as problematic

---

[2] Don R. Ostrander to Wernher von Braun, 20 March 1961, Centaur files. See also Ostrander to von Braun, 29 March 1961, John Sloop Papers, NASA Historical Reference Collection.

[3] Abe Silverstein to Associate Administrator, 6 September 1961, papers of John Sloop, NASA Historical Reference Collection.

as he had heard from various sources, including Donald Heaton. A meeting attended by Administrator James Webb, Silverstein, von Braun, Nicks, Ostrander, and Edgar Cortright, Assistant Director of Lunar and Planetary Programs, confirmed his worst fears. Nicks recalled, "The meeting produced a formal position by Wernher von Braun that the future of Centaur was totally uncertain. We left the meeting with the clear understanding that our Centaur-based planetary missions were postponed indefinitely."[4]

Originally, the payload capability of Centaur was set at 2,700 to 2,800 pounds for lunar missions. Based on the projected weight of Centaur, JPL's contract with Hughes for Surveyor called for a spacecraft weight of 2,500 pounds with a science payload of approximately 340 pounds. In the spring of 1962, Marshall managers informed JPL that Centaur could not lift this weight and asked for a reduction in payload weight to 2,100 pounds. This weight change had a demoralizing effect on JPL and Hughes engineers because they had to redesign the Surveyor mission to accommodate fewer scientific experiments.

Whether Surveyor was even necessary for the Apollo program was called into question in 1962. Nicks recalls in his personal memoir *Far Travelers* that Max Faget, one of the leading engineers on the Apollo program, gained the ear of Congressman Joseph Karth (D-Minnesota), the powerful head of the Subcommittee on Space Sciences of the U.S. House of Representatives Committee on Science and Astronautics. Nicks claims that Faget flatly declared that not only was Surveyor superfluous as a prototype for the Lunar Excursion Module, but even the data Surveyor was supposed to provide on the topography of the Moon could be obtained by other means. Faget's meddling with a powerful Congressman raised a furor at Headquarters. "Webb promptly set the record straight about Surveyor's importance to Apollo," Nicks relates, but Apollo engineers continued to believe they would never have data they needed from Surveyor. Although they were careful not to speak openly of doubts about Ranger and Surveyor, Nicks believes they were convinced that they could not depend on these questionable programs.[5]

Apollo designers may have refrained from directly attacking Centaur, but Karth felt no such compunction. Congressional hearings in March tackled the issue of the escalating costs of NASA's programs. Congress wanted to terminate the Centaur and Surveyor programs. Karth thought NASA should admit its mistake in developing Centaur. He was especially critical of the fuel-feed system used in the RL10 engine. But NASA and the Air Force presented a unified front in defense of Centaur.[6] A decade later, policy analyst Erasmus Kloman speculated that although scientists had probably been fairly sure of the composition of the Moon's surface by the time Surveyor was launched, it "provided a forcing mechanism for the development of Centaur as

---

[4] Oran Nicks, *Far Travelers*, 17.

[5] Ibid, 91–92.

[6] Brian Duff, "NASA Officials Defend Centaur," *San Diego Union* (20 March 1962).

part of the space agency's long-range launch vehicle program."[7] NASA evidently thought the space program needed liquid-hydrogen fuel and that Centaur would prove its feasibility.

By December, the Centaur program was in crisis. With Atlas-Centaur already installed on the launch pad at the Cape, an extremely serious new problem had surfaced. An ambitious test program run at the company's Sycamore Canyon facility revealed leaks in the intermediate bulkhead that separated the liquid-hydrogen and liquid-oxygen tanks. Hydrogen was seeping through imperceptible holes in the welds. These holes seemed to open up only after the liquid hydrogen was loaded into the tank. This hydrogen leakage was a serious problem because it destroyed the vacuum between the oxygen and hydrogen tanks and increased the transfer of heat across the bulkhead. The leaks called into question the basic design of the rocket. General Dynamics asked some of its best engineers to tackle the problem, including Atlas designer Charlie Bossart.

Headquarters sent John Sloop, then NASA's Deputy Director of Launch Vehicles and Propulsion Programs, to investigate exactly what was happening at the company. Sloop, a strong advocate of high-energy fuels, had worked in the rocket section at Lewis Research Center until called to Washington by Silverstein. In a confidential memo to Homer Newell, written in December 1961 after a visit to General Dynamics, Sloop revealed that the company considered the problem of leaks across the bulkhead so serious that the integral tank design might have to be scrapped. He reported that the company thought that it might still be feasible to fly the Surveyor and Mariner missions on Centaur because the hydrogen leaked so slowly. But because of the longer coast time of the Advent mission, liquid-hydrogen leakage might jeopardize the mission. "[General Dynamics/Astronautics] believes that the only safe way to meet all Centaur missions is to drop the integral tank design and to go separate fuel and oxidizer tanks,"[8] Sloop noted. This was a blow to the defenders of the integral tank design, particularly Charlie Bossart. The memo noted that Marshall had already requested General Dynamics to begin a design study using separate tanks. Sloop argued against abandoning the innovative tank design, urging a redesign of the bulkhead with greater wall thickness at the points of welding. Whether this would provide a real solution to the problem, however, was not yet clear.

Sloop's second report, filed a few days later, contained an indictment of General Dynamics' management structure. At this time General Dynamics had a matrix structure with all the projects drawing support from the operating divisions. Most of its 32,500 employees worked on the Atlas project, which commanded the lion's share of the company resources. Ehricke was one of five program directors. He had only twenty-seven people assigned to the project, only five of whom reported directly to him. Deane Davis, the Centaur Project Engineer, had two that he directly supervised. Although the men assigned to Centaur had specialized knowledge of Centaur

---

[7] Erasmus Kloman, *Unmanned Space Project Management: Surveyor and Lunar Orbiter* (NASA SP-4901, 1972), 10.

[8] John Sloop, "Memorandum for Director of Space Sciences," 18 December 1961, John Sloop papers, NASA Historical Reference Collection.

components and subsystems, only Ehricke and Davis were concerned with overall performance and systems. Sloop observed, "Mr. Ehricke is doing a Herculean task but Centaur is too big for one man. What is badly needed is strength in depth; namely, give Mr. Ehricke a larger staff directly under his control."[9] The small team of five Marshall engineers who were monitoring the contract in the Resident Office were powerless to effect the sweeping organizational changes that were needed to make Centaur a viable program.

At last, NASA acted to exert more direct control over the Centaur program. Marshall Space Flight Center set up a program office in Huntsville to provide greater oversight. In response to pressure from NASA, General Dynamics/Astronautics President James Dempsey directed Grant Hansen to replace Krafft Ehricke as head of the Centaur program. Krafft Ehricke became Director of Advanced Studies, charged with the company's long-range plans and applications for Centaur hardware. Dempsey promoted Hansen to vice-president reporting directly to him, bypassing the chief engineer of the company, Mort Rosenbaum, who had been at loggerheads with Ehricke. In discussing the transition in leadership, Hansen explained, "And then finally came sort of a showdown and, as Dempsey explained it to me, they felt that Krafft was a tremendously imaginative, creative idea man, a hell of a good engineer, but that he wasn't enough of an S.O.B. to manage a program like this. He had functional department people who would tell him to get lost, and he would be willing to do just that, and they needed somebody who wouldn't."[10] Hansen proved to be a skilled manager who possessed the authority and resources that Ehricke had lacked. The early months of his tenure would be marked by a major reorganization of the project, the failure of the first flight, and a highly critical congressional investigation. Undeterred, Hansen headed Centaur through the end of the Surveyor program.

Born in Bancroft, Idaho, in 1921, Hansen had been brought up in California, where he attended San Bernardino High School and Junior College. He had served in the Navy during World War II as an electronics technician and engineer. The G.I. Bill made it possible for him to earn a B.S. in electrical engineering at Illinois Institute of Technology. He did graduate work at the California Institute of Technology while he worked in missiles and space systems engineering with the Douglas Aircraft Company from 1948 to 1960. At Douglas, he designed the Nike launch control and test equipment and supervised test firings at White Sands Proving Ground. He had responsibility for electrical systems for Nike Ajax, Nike Hercules, Honest John, Sparrow, MB-1, Thor, Thor-Delta, Nike Zeus, and Skybolt missiles. Moving from Douglas to General Dynamics in May 1960, Hansen became Chief Design Engineer, in

---

[9] John Sloop, "Memorandum for Director of Space Sciences, 20 December 1961, John Sloop papers, NASA Historical Reference Collection.

[10] John Sloop interview with members of the Aerospace Division Convair/General Dynamics, 29 April 1974, NASA Historical Reference Collection.

charge of all design and laboratory testing, and also became the production-engineering liaison on the Atlas and Centaur programs.

Hansen had known that Centaur was in trouble, but the promotion took him by surprise. He read it in a memo that simply showed up in his mailbox.[11] His appointment as vice-president in charge of the Centaur program coincided with the change in management structure from a matrix to a project-type organization. In announcing the change to employees, a company newsletter stated, "In arriving at the Project Centaur decision, two principal factors were considered by management: the future of the company demands to a large degree on upgrading the Centaur effort to an exceptional level of efficiency, and also the company must utilize Ehricke's unique creative talents to the fullest degree possible for originating new products, just as he originated the Centaur."[12] Hansen's integrity and calm leadership style earned the respect of those with whom he worked, both inside the company and within NASA.

The reorganization, called "projectization," gave Hansen more direct control and freedom to build an organization from the ground up. "We had some very good engineers," he said. "I could hand-pick the best ones to be part of the Centaur project, and I did that. So we had people like Red Lightbown and many others who were the best engineers that Convair had."[13] General Dynamics assigned 1,000 employees to the project. They took over a building vacated by one of the Atlas groups. To assist Hansen in implementing the change to a project structure, the company hired a consulting firm, Robert Heller and Associates.

In a letter to Marshall program director Hans Hueter, Jim Dempsey emphasized how the reorganization would improve Centaur operations: "Such an organization change cannot of itself automatically solve all the problems, but can serve as an effective tool for improving attention to the program, facilitating effective internal communications, clarifying responsibility and authority, speeding up program activity, and identifying team relationships and appropriate communication channels with the Marshall Space Flight Center and the General Dynamics/NASA office." Dempsey assured Hueter that the new launch date would not be delayed by this reorganization.[14]

In an appended memo to Hueter, Dempsey assured him that Atlas expertise would be available to Centaur. The Atlas launch team at the Atlantic Missile Range would not only be responsible for Atlas, but would also act as a subcontractor for Centaur. Dempsey further assured Hueter that engineers with Atlas knowledge and experience were being transferred to Centaur. In addition, Hansen would participate in Dempsey's daily staff meeting. Atlas reports

---

[11] Interview with Grant Hansen (former employee of General Dynamics) by Virginia Dawson, 6 June 2000.

[12] "FYI: Reorganization for Project Centaur," 29 January 1962, John Sloop papers, NASA Historical Reference Collection.

[13] Interview with Grant Hansen by Virginia Dawson, 6 June 2000.

[14] J. R. Dempsey to Hans Hueter, 12 February 1962, John Sloop papers, NASA Historical Reference Collection.

would be distributed to Centaur personnel, and Centaur personnel would participate in the Atlas engineering change board reviews. Dempsey acknowledged that one of the weaknesses of a project organization, as opposed to one that was organized in functional groups, was its isolation within the company. He assured Hueter that Centaur top management would make a special effort to stress a "cross feed of learning." He wanted Marshall representatives to "have maximum visibility into [General Dynamics/Astronautics] Centaur operations so that they have every opportunity to aid in ferreting problems."[15]

To implement more rational schedule planning and control, NASA insisted that General Dynamics adopt Program Evaluation and Review Technique (PERT) and Companion Cost reporting. PERT, a time-/cost-management tool first used by the Navy to develop the Polaris missile, was adopted by NASA in late 1961 to impose discipline on the management of large projects by developing a master schedule for every task of the program and tracking progress. Each department head was responsible for accomplishing a specified task within an allotted timeframe. Dempsey assured Hueter, "We have adopted the attitude that any schedule slippages not directly assignable to a[n] unforeseeable technical development problem are indicators of lack of competence on the part of either the Schedule Planning and Control personnel or the responsible function line managers."[16]

Dempsey expected everyone in the organization to think in terms of systems engineering. "Each part must be designed, each procedure written, each test planned and accomplished with full recognition of how that effort fits into the total picture and with concern for the implications upon all other elements of the system."[17] To strengthen systems engineering at General Dynamics, Deane Davis was given more control. As Deputy Program Director for Technical Control, he was responsible for all technical planning and integration. At the same time, design engineering under William Radcliffe was broadened to include system design engineering. Centaur design personnel drawn from design groups throughout the company were organized into nine specialized design groups, such as propulsion, structures, and guidance. Hansen believed that the move from matrix to project management was just in time. Without this reorganization, the program would have been canceled. "I knew that I had to get the problem solved, and get it working successfully, or NASA was going to slam the door on us," he said.[18]

But the change from matrix to project management without any real effort to control the program by Marshall could not solve all of Centaur's problems. In Hansen's view, von Braun

---

[15] Ibid.

[16] Ibid.

[17] Attachment to letter to H. Hueter from J. R. Dempsey, 12 February 1962, John Sloop papers, NASA Historical Reference Collection.

[18] Interview with Grant Hansen by Virginia Dawson, 6 June 2000.

"just sort of pushed it into a backroom and let it be handled by Hans Hueter." Hueter had served as Chief Test Engineer at Peenemünde, emigrating to the United States with the other members of the von Braun team in 1945. His German background did not prepare him for the entirely different engineering philosophy at General Dynamics. The two cultures inevitably clashed. From Deane Davis's perspective:

> To the Huntsville people the methods and philosophies of the Atlas and Centaur were as mysterious as the dark side of the Moon. On the Astronautics side, still riding the crest of their brilliant success with the Atlas, and with no reason to believe that the Centaur would prove to be any different, the new Centaur people resented what they felt was undue interference in their established manner of creating, producing, and operating their product. After all, they said repeatedly, the Atlas and Centaur were their inventions.[19]

Unlike most of the Germans on the project, Hueter was actually loved and respected by his "unruly wards at Astronautics." Hueter had the impossible task of reconciling what Davis considered to be two diametrically opposed design and administrative approaches. Shortly after the program was transferred to Marshall, the Convair plant in San Diego was "deluged with Huntsville personnel attempting to understand their new creature." Apparently, whenever the Germans came to San Diego, Atlas designer Charlie Bossart always tried to keep a low profile. During one visit, the Germans took a tour of the plant and visited the Sycamore Canyon hot-firing testing facility. Then they all sat down for a late-day briefing. Trouble began when Bossart and Willie Mrazek, von Braun's Structural Section Chief, began to argue in loud whispers. Their voices began to rise over the subject of the rocket's structure. "Charlie was trying to explain its merits to a disbelieving Mrazek and from experience I knew that the only solution was to get those two gentlemen separated from the briefings so they could have at it."[20]

Bossart led Mrazek out into the factory yard, where a Centaur tank stood gleaming in the sunlight. Mrazek asked, "What's inside it?" To which Bossart responded, "Nitrogen." Nitrogen was used for pressurization until the rocket was filled with its liquid-hydrogen/liquid-oxygen propellants just prior to launch. Without pressurization, the thin skin of the rocket would first wrinkle, then collapse. Mrazek, familiar with the solid, reinforced walls of rockets designed by the von Braun team, was perplexed by Bossart's insistence that nitrogen, kept at the relatively low pressure of 8 to 10 pounds per square inch, was sufficient to keep the tank rigid.

He also questioned how such a thin, unbuttressed structure could be strong enough to carry a rocket aloft. To quell Mrazek's doubts, Bossart invited him to take a sledge hammer and give the tank a whack. Failing to put even the slightest dent in the tank, he tried again, this time giving the

---

[19] Deane Davis, "Seeing Is Believing, or, How the Atlas Rocket Hit Back," *Spaceflight* 25 (5 May 1983): 196–198.

[20] Ibid.

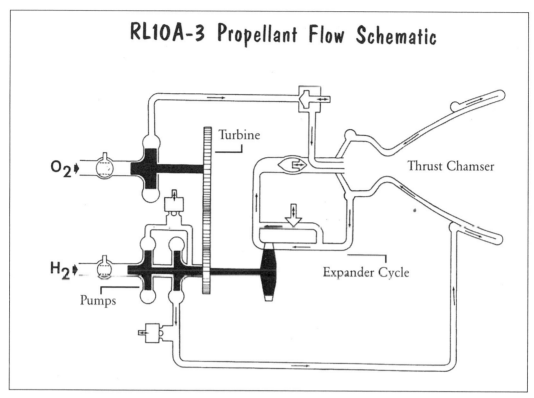

Diagram of RL10 engine: Liquid oxygen and liquid hydrogen are pumped through tubes toward the engine's thrust chamber. Liquid hydrogen cools the walls of the thrust chamber while it picks up heat and becomes gaseous. The engine's unique "expander cycle," in which a small amount of gaseous hydrogen in the exhaust stream flows backward to drive the pumps of turbine, can be seen in the center of the diagram. (Courtesy of Pratt & Whitney)

side of the tank a glancing blow that caused the sledge hammer to fly out of his hand, knocking his glasses off, but again leaving the surface unscathed. Although this test may have proved the strength of the balloon structure, it did nothing to endear General Dynamics to Mrazek or win the von Braun group's faith in the ability of Centaur to lift an expensive spacecraft into space.[21]

Shortly after the reorganization of the program, the Aerophysics Group at General Dynamics began to tackle the aerodynamics of the integration of the two rockets. This was a major undertaking that Don Lesney realized would be of enormous importance to the success of Centaur. In addition to a big interstage adapter to connect Atlas with Centaur, the vehicle carried a 10-foot-diameter payload fairing. What concerned Lesney was whether the middle of the vehicle was strong enough to withstand the aerodynamic stresses caused by flying through the wind. The vehicle was double the length of the original Atlas ICBM. Questions such as how thick to make the skin on Atlas and

[21] Ibid.

how much pressurization was necessary to keep it structurally intact could only be solved through testing. Lesney recalled that because his Huntsville counterparts never returned his calls, he was unable to get the go-ahead for wind-tunnel studies.[22] Management of RL10 engine development by the von Braun group produced similar tensions, although it saw the potential of the RL10 as the propulsion system for one of Saturn's upper stages. In contrast, it seemed to want little to do with the upper stage launch vehicle for which the RL10 was originally designed.

Troubles at Pratt & Whitney

Pratt & Whitney's engineers had developed confidence in handling liquid hydrogen fuel during Project Suntan. They relished the prospect of contributing this new expertise to the space program. However, Pratt & Whitney was a turbojet engine manufacturer and had no prior experience in designing a rocket engine. The company had signed a contract with the Air Force in mid-October 1958 to develop the RL10 engine—an engine whose innovative design and outstanding reliability would earn the designation of historic engineering landmark from the American Society of Mechanical Engineers in 1979. Since the 304 turbojet engine they had designed for Suntan was air-breathing, the first hurdle was the choice of oxidizer. Originally, the company's engineers favored fluorine because of its superior efficiency, but fluorine proved so corrosive and volatile that they settled on liquid oxygen. The turbopump designed for the 304 dictated the RL10's original 15,000-pound thrust rating.[23]

Initially, Pratt & Whitney engineers did not try to adapt the unique expander cycle of the 304 to the RL10. Early designs, based on the conventional gas generator cycle, required a separate combustion chamber to start the turbopumps. A suggestion from Perry Pratt, the company's chief engineer, led the design team back to the elegant simplicity of the 304. Instead of fighting against the inherent properties of liquid hydrogen, the expander cycle made them intrinsic to the design.[24]

The engine was designed and built at the company's factory in East Hartford, Connecticut. Testing took place at the company's new Florida Research and Development Center. The RL10 test team reveled in the freedom from interference it enjoyed in this remote swampland near West Palm Beach. By July 1959, they had completed 230 successful firings in a horizontal test stand. The RL10 started, stopped, and restarted with surprising reliability. By November, the technical problems appeared to be nearly solved—not so the escalating costs of developing the new engine.

Prior to contracting with NASA, Pratt & Whitney government contracts had always been fixed-price. With this type of contract, any cost overruns had to be assumed by the company. The

---

[22] Interview with Don Lesney, 5 June 2000.

[23] Joel E. Tucker, "The History of the RL 10 Upper-Stage Rocket Engine, 1956–1980," published in *History of Liquid Rocket Engine Development in the United States, 1955–1980*, ed. Stephen E. Doyle, AAS History Series, vol. 13 (San Diego: AAS, 1992), 123–151.

[24] Mulready, *Advanced Engine Development*, 60.

NASA engineers Ali Mansour and Ned Hannum review test results of Pratt & Whitney's RL10 engine at Lewis Research Center, 1963. (NASA C-64329)

cost-plus-fixed-fee contract for the RL10 allowed the company to bill for additional development costs when they ran into unforeseen problems. Initially, Mulready apparently neglected to submit all the design changes to NASA, simply addressing the problems as they arose. "We had not learned to wait and put changes into the contract. It was noticed that Convair, which was an old hand at the cost-plus business, tended to 'keep its head down and point left' in times of trouble."[25] Once Pratt & Whitney began to bill for design changes, costs escalated. Mulready recalled how after a dressing-down at NASA Headquarters, he returned to Florida determined to cut costs. Confident after one hundred successful test firings that the engine was extremely reliable, he canceled test equipment for the ignition system—a decision he later regretted. Only one engineer challenged him, claiming that data revealed that the "engine was not lighting on the same spark every time."[26] Mulready dismissed this objection as ludicrous.

Meanwhile, Lewis Research Center became deeply involved in the development of the RL10 engine. Lewis had previously developed a regeneratively cooled fuel/oxidizer injector that served as the basis for the engine's injector design. Pratt & Whitney engineers added a textured surface of special aluminum mesh called Rigi-Mesh that facilitated cooling.[27] NASA Headquarters not only asked Lewis engineers to assist Pratt & Whitney in troubleshooting, but also wanted them to develop an in-house competency that would be useful in monitoring the work of the contractor. In March 1960, the Center initiated a 10-month test program with three RL10 engines. In addition to providing a check on contractor performance, this testing yielded a complete altitude evaluation of the propulsion system.[28] By November, Lewis engineers had completed two successful firings of the engine in the altitude wind tunnel. In recognizing the contributions of Lewis Research Center, Ostrander wrote, "The Lewis staff assistance, which is based on an extensive technical base and on unique experience in handling these propellants in similar equipment, has been and is invaluable."[29]

After a series of tests of the engine in the horizontal position, Pratt & Whitney initiated tests of two engines in October 1960 in a new vertical test stand called E-5 at the company's Florida test center. The engine would actually be flown in the vertical position, so these new tests were especially important. The first test went as planned, but during the second test, the engine exploded. Investigation suggested that the explosion could be attributed to a human or procedural error. Nevertheless, the mishap caused the Centaur launch schedule to slip to

---

[25] Mulready, 76.

[26] Ibid.

[27] Mulready, 70. See also Sloop, *Liquid Hydrogen*, 191–193; Bilstein, *Stages to Saturn*, 134–140; Tucker, "History of RL 10," 132.

[28] Centaur Propulsion System Testing, Contract NAS9-2691, Pratt & Whitney, 7 December 1962, Vol. 1, Box 59, Old RL10 Records (Goette files), DEB vault, NASA Glenn.

[29] General Don R. Ostrander to Eugene Mangeniello, 23 June 1961, Box 254, RG 255, NARA, Chicago, IL.

20 December 1961. At this point, the engine also had thrust-control problems, in addition to hydrogen leakage rates in excess of specifications. Pratt & Whitney fixed the damaged test stand and resumed testing on 12 January 1961.

After the engine exploded a second time during an important demonstration to Air Force and NASA officials in the spring of 1961, Mulready immediately ordered the igniter test rig he had previously canceled. Testing determined that in the horizontal position, a small amount of liquid oxygen, previously unnoticed, flowed back toward the spark, mixing with the hydrogen gas. In the horizontal position, this facilitated ignition, but in the vertical position, gravity cut down the backflow of oxygen, thereby preventing the firing of the engine. The installation of an oxygen feed line to the igniter solved the problem. On 24 April 1961, two RL10 engines successfully completed their first full-duration test on the E-5 test stand.

Von Braun used the two engine explosions as his cue to take a more assertive role in management of the contractor. (He called this "penetrating" the contractor.) Marshall engine specialist Leonard C. Bostwick insisted on regular meetings and a formal development plan– something the fiercely independent Pratt & Whitney engineers found disruptive and inappropriate. Mulready wrote, "The rapid growth of the RL10 technology, which was totally new in so many dimensions, could not have happened with the formality that NASA sought."[30]

But working with NASA proved to be a positive experience for Pratt & Whitney in many ways. Although its engineers were enormously competent turbojet designers, they benefited from the rocket expertise of the Marshall and Lewis teams. For example, at Pratt & Whitney, turbojet engines were always given a manual countdown. NASA insisted on automated test stands. "The gentle nature of the RL10, which operates similarly to a jet engine, was the only reason that Pratt & Whitney was able to use the old system as long as it did," Mulready admitted.[31]

In November of 1961, the Air Force contracts with Pratt & Whitney were transferred to Marshall. By May of 1962, the RL10 engine had completed over 700 hot firings. It proved to be so reliable that Marshall chose to power the Saturn S-IV stage with six RL10 engines. When called to testify at the Centaur hearings in May 1962, Bruce Torel asserted proudly that all of the development problems of the "free world's first liquid hydrogen rocket engine" had been solved.[32] He was correct, although Pratt & Whitney would have to wait more than a year for General Dynamics to solve many of Centaur's technical problems before the RL10 would fly.

---

[30] Mulready, 83.

[31] Ibid.

[32] Centaur Program, 1962 Hearings, 111

Launch Operations at the Cape

At Cape Canaveral, launch operations for unpiloted vehicles were located at the Atlantic Missile Range east of the Banana River. This area under Air Force control originally contained one launch pad for Atlas-Centaur, referred to as Pad 36A. In addition to the first pad, a second pad, 36B, was completed in the mid-1960s. A Central Control Building, situated a safe distance from the launch sites, housed the Air Force range safety officers. They were responsible for all range support, including destroying a vehicle if it suddenly veered off course.

Initially, Marshall managed Atlas-Agena, Atlas-Centaur, and Thor-Agena launch operations at the Cape. This included all robotic scientific and applications missions except Delta, which was under the control of Goddard Space Flight Center. Goddard personnel at the Cape were all former Vanguard rocket people. General Dynamics was responsible for integrating Centaur with Atlas, while JPL was responsible for integrating the spacecraft with the launch vehicle.

Building AE, previously a staging area for Air Force missiles, played an important role in prelaunch operations. Oran Nicks had the former hangar air-conditioned. Nicks remarked, "While prelaunch checkout facilities steadily became less ramshackle, it was 1964 before we could begin to treat our interplanetary travelers with the care they deserved."[33] A cleanroom was built onto the rear of the building that was later used for the installation of Centaur shrouds.

Everyone at the Cape during the early NASA years was learning. James Womack, one of the first NASA staff at the Cape, recalled that he was familiar with more conventional propellants but had never worked with liquid hydrogen. Because the Cape operations staff knew little about liquid hydrogen, they attended the required safety lectures. They were cautioned that since hydrogen produces no visible flames during combustion, they should take a broom along every time they went near a hydrogen line. If the broom did not burst into flames, it was considered safe to proceed. They quickly realized that this precaution was not necessary; although liquid hydrogen required attentiveness and respect, it was considerably less volatile than they had been led to believe.[34]

Roger Lynch was one of the first Convair employees assigned to Cape Canaveral in early 1961. A graduate of the U.S. Merchant Marine Academy who had joined General Dynamics in 1956, Lynch's first job at the company was at Sycamore Canyon. He was asked by Ken Newton to go to the Cape to ready the Atlas-Agena launch site in preparation for the launch of the Ranger series of Moon shots. He was transferred to Pad 36 in anticipation of the launch of the first Atlas-Centaur.

During that time, Lynch observed the early maneuverings of the Marshall Space Flight Center engineers as they began to make their presence felt at the Cape. They were in the process of installing the large Saturn bureaucracy on the other side of the Banana River (at the site later to be known as Kennedy Space Center) under Kurt Debus. Relations with the Air Force and

---

[33] Oran Nicks, *Far Travelers*, 82.

[34] Interview with Jim Womack by Virginia Dawson, 11 November 1999.

Wernher von Braun anxiously monitors the launch of a Pioneer 4 spececraft for a lunar flyby using a Juno II, 1959. (NASA 9131491)

Marshall were strained because the Air Force was determined to maintain control of robotic operations. General Dynamics managed a launch site under the distant supervision of Marshall personnel. Initially, the Marshall people did not take much interest in the preparations at Pad 36. As the date of the first launch approached, Roger Lynch became aware of the contempt of Marshall engineers for the pressure-stabilized design of Atlas and Centaur.

> We had come out of a very successful major ballistic weapons system program with technology that was foreign to Marshall. Our monocoque structure was considered high risk. They took exception to all aspects of the vehicle design. It was different in every way possible from what they were starting to design for Saturn. I was in a meeting at Huntsville when it was admitted that even Saturn, for all of its structure, would collapse in flight without internal tank pressure. Over drinks, they admitted that it made little sense to fly all that structure. The real reason for their design was a passion for vertical check-out.[35]

---

[35] Letter from Roger C. Lynch to Virginia Dawson, dated 4 June 2000.

Lynch observed that Marshall engineers were equally conservative with respect to electrical systems. They preferred old-fashioned relays and vacuum tubes. Their junction boxes were so heavy that he thought they belonged on a battleship, not a spaceship. General Dynamics and Honeywell designers were pushing the state of the art with much lighter solid-state electronics, although components often proved unreliable.

The von Braun team was also critical of the ground support equipment used by General Dynamics. At this time, Atlas and Centaur were considered two distinct systems with a minimum interface. Servicing the two systems was more complicated than it would have been if an integrated system had been designed from scratch. A whole new vacuum technology that included the design of large valves and transfer lines had to be developed for liquid hydrogen. General Dynamics engineer Denny Huber recalled that the design of the vacuum insulation posed quality-control problems and limited the number of suppliers willing to bid on vacuum-system hardware. "If a potential supplier had no capability to handle liquid hydrogen, his design approach was at risk."[36]

Because it was the first time the two vehicles had been mated together, the General Dynamics launch team was inundated with "fierce and unpredictable" design changes.[37] This was intolerable to the Huntsville team, which worked from more detailed specifications and left little room for adjustments by the contractor at the launch site. Von Braun may have been referring to the chaotic situation at Pad 36 when he observed that it was "better to build a rocket in the factory than on the launch pad."[38]

The Failure of F-1, the First Atlas-Centaur Launch

Liquid-hydrogen leakage, in addition to guidance and engine problems, kept Atlas-Centaur sitting like a lone sentinel over the Cape's broad expanse of beach from October 1961 through the following spring. Finally, it was ready. On 8 May, F-1 lifted off into the clear Florida skies for its maiden voyage. As an eyewitness in the crowded blockhouse reported, "While the elated engineers were shaking hands and congratulating each other the vehicle reached plus 54 seconds into the mission, and a sudden yell of 'Missile blew up! Missile blew up!' came from one of the observers at the periscopes."[39] Two seconds later, the vehicle was completely obscured by a massive fireball fueled by thousands of gallons of liquid hydrogen and kerosene fuel. NASA quickly determined that Centaur was the cause. Aerodynamic pressure on the cover protecting the insulation (called the weather shield at that time) had caused it to burst, ripping away the

[36] E-mail communication from Denny Huber to Ed Bock, 30 March 2002.

[37] Letter from Roger C. Lynch to Virginia Dawson, dated 4 June 2000.

[38] Andrew J. Dunar and Stephen P. Waring, *Power to Explore: A History of Marshall Space Flight Center 1960–1990* (Washington, DC: NASA SP-4313, 1999), 39–51.

[39] Joseph Green and Fuller C. Jones, 8–41.

insulation and exposing the walls of the fuel tank to the heat of the atmosphere. Pressure buildup in the tank from the boiling off of the liquid hydrogen caused the fuel tank to rupture, spilling volatile liquid hydrogen down the sides of the rocket, where it was ignited by a spark from the engine. The loss of Centaur was a serious setback for NASA.

The aborted first flight of Centaur put NASA on the defensive. The House Subcommittee on Space Sciences, headed by Joseph Karth (D-Minnesota), called for a probe. Karth questioned whether Centaur was a key to the nation's future rocket propulsion needs or simply an increasingly expensive national liability. The troubled Centaur program had implications for national security. Four and one-half years after Sputnik, American rocket propulsion technology still lagged behind that of the Soviet Union.[40] Centaur had been billed as an important new propulsion technology. Its failure reinforced the perception of American technical inferiority. Now, with Centaur considered the key to a soft landing on the Moon, the subcommittee needed reassurance that the space program was still on track. Spokesmen for General Dynamics, Pratt & Whitney, and Marshall Space Flight Center knew that this was by no means a routine investigation. The fate of the Centaur program hung in the balance.

Given the enormous technical and managerial challenges involved in developing this new technology, the failure of F-1 seemed almost inevitable. However, the timing was unfortunate. The Soviet Union had already succeeded in crash-landing two lunar vehicles on the Moon, while five attempts by the United States Ranger vehicles had all failed. Both Ranger and Surveyor were needed to provide scientific data on radiation, magnetic fields surrounding the Moon, and the Moon's surface topography. The Ranger failures, compounded by the aborted flight of F-1, forced NASA to consider whether the beleaguered Centaur program should be canceled.

Wernher von Braun and Hans Hueter admitted that Marshall had inadequately supervised General Dynamics, but they blamed the laissez-faire approach of the Air Force, which had allowed the program to evolve without adequate oversight. Von Braun thought that in-house competence by the government and complete control over contractors was the only way to avoid development problems. He stated:

> I think the general lesson we have learned from Centaur management is probably that one should never underestimate the magnitude of a program where so many new and unproven ideas are tried out and I think we will always get in difficulties, as a Government agency, unless we build up a competence in the Government that we can really stay on top of the problems right from the outset and learn how to identify potential problem areas before we have explosions and fires and setbacks.[41]

---

[40] Walter LaFeber, *America, Russia, and the Cold War, 1945–1996*, 8th edition (New York: The McGraw-Hill Companies, Inc., 1997), p. 192.

[41] Centaur Program, 1962 Hearings, 59.

Marshall had taken over management from the Air Force after many of the important design decisions had already been made. Von Braun admitted that they had not "penetrated" the program sufficiently. "The only excuse one can have for it is that it started out as a little exploratory program and grew and grew and grew into a major program, and it wasn't intended that way from the outset."[42]

Marshall's approach to contracting with industry differed radically from that of the Air Force. Von Braun was used to designing, building, and testing a rocket prototype before turning it over to an industry contractor for closely supervised production.[43] The "everything-under-one-roof approach" had prevailed at Peenemünde. It dovetailed with the U.S. Army arsenal tradition of weapons development. Von Braun believed that to manage contractors properly meant exerting control over every aspect of testing and manufacture. Marshall furnished its contractors with detailed design requirements and monitored their operations closely, assigning as much as 10 percent of its staff in the 1960s to resident offices.[44] In contrast, the Air Force, because it lacked extensive in-house expertise, allowed its industry contractors to develop designs and independently work out solutions to problems.

But more than management bothered von Braun. He had serious reservations about the basic design of Centaur. He made it clear that Marshall Space Flight Center would never have tolerated the balloon structure. He called the approach that General Dynamics had taken imaginative, but risky. "In order to save a few pounds, they have elected to use some rather, shall we say, marginal solutions where you are bound to buy a few headaches before you get it over with. Ultimately when you are successful you have a real advanced solution." But von Braun did not believe that Centaur would ever be successful. Referring specifically to the Centaur pressurized tank, he said, "It is a great weightsaver, but it is also a continuous pain in the neck."[45] He pointedly declared that industry contractors would not have the same design freedom when it came to Saturn development. "We are really making an all-out effort to stick together with the contractor before major design decisions are even made and have our men argue with his men as to whether this is really the way to go."[46]

Krafft Ehricke and Grant Hansen defended the technical decisions that determined rocket structure and fuel choice. A modest person without the aristocratic pretensions and confidence of von Braun, Hansen admitted that he was rather frightened when called to

---

[42] Ibid.

[43] On the arsenal approach, see Michael J. Neufeld, *The Rocket and the Reich: Peenemünde and the Coming of the Ballistic Missile Era* (New York: The Free Press, 1995), 108.

[44] Andrew Waring, *Power to Explore*, 39–51.

[45] Centaur Program, 1962 Hearings, 58–59.

[46] Ibid., 59.

testify. But he said to himself, "All I've got to do is tell the truth, and convince them that we're understanding the problem and are working it and it should go away."[47] Given the cost overruns and delays, he found the legislators' concerns to be justified: "When a program is that far behind schedule and costs that much money and blows up on the launch pad, they owe it to the public to hold a hearing to find out what's going on, and whether or not these programs should be continued or not." Asked at the hearing about the difference in design philosophy between General Dynamics and Huntsville, Hansen agreed that Huntsville's was more conservative. In contrast to the 35-percent margin of safety the von Braun team specified in its contracts, the contract between General Dynamics and the Air Force allowed a 25-percent margin. He explained, "We are inclined, I think, to be willing to take a little bit more of a design gamble to achieve a significant improvement, whereas I think they [Huntsville] build somewhat more conservatively."[48] To designers at General Dynamics, the pressure-stabilized tank may have been a "design gamble," but its advantages made it one worth taking.

Although no longer in charge of the program, Krafft Ehricke vigorously defended the design, pointing out that the pressure-stabilized tank was also used on Atlas, a rocket deemed reliable enough to send Mercury astronaut John Glenn into orbit around Earth. In Centaur, the pressure-stabilized tank design optimized the unique characteristics of liquid hydrogen—its low density and extreme cold. The result was a dramatic savings in weight, always the driving factor behind any successful rocket design. Ehricke also strongly defended the use of liquid-hydrogen fuel. He admitted that hydrogen leakage through minute holes in the welds had proved a problem, but not an intractable one: "We have definitely in this particular case, if I might say so, gambled at a very low weight and found that we have to correct ourselves."[49] At another point in his testimony, he said, "Hydrogen itself has turned out to be less of a culprit than many thought initially. Hydrogen, like all chemical fluids behaves fine if you know its little 'idiosyncrasies' and treat it correctly. But you have to go through a development program such as ours first."[50]

To Ehricke, Centaur provided an example of a transition from the development of a strategic weapon to a launch vehicle for the peaceful exploration of space.[51] A launch vehicle required far greater thrust and reliability than a missile. He pointed out that in addition to the pioneering work by General Dynamics on liquid hydrogen, the company was breaking

---

[47] Interview with Grant Hansen by Virginia Dawson, 6 June 2000.

[48] Centaur Program, 1962 Hearings, 95.

[49] Ibid., 97.

[50] Ibid., 69.

[51] Ibid., 67–68.

new ground in important technical areas, such as restarting the vehicle after an extended coast in zero gravity. Centaur also represented a transition stage in government contracting. Centaur's management problems had proved that the Air Force's hands-off model of contract management was not appropriate for a civilian agency.

Ehricke thought that many of the troubles with Centaur could be attributed to the fact that it had never been accorded the national priority classification it deserved. Centaur was just as important to the national space program as the piloted space programs—Mercury, Saturn and Apollo—which all had DX priority (the highest national priority in the procurement of materials). Indeed, President Kennedy's announcement in May 1961 that the nation would put a man on the Moon within a decade made it imperative that Centaur succeed in sending Surveyor to soft-land on the Moon. Lack of DX priority had made it more difficult to obtain parts from contractors in a timely manner. Ehricke said, "Like other space programs, Centaur was to grow as it went along. However, unlike Mercury and Saturn, Centaur could never attain DX priority although its contributions are as much a cornerstone of our future space capability as those of a manned space capsule and a high thrust booster."[52] The other constraint was the limitation imposed on Centaur by the requirement that it be designed using off-the-shelf parts that belonged to Atlas. Although this saved money, it led to design compromises.

After the Centaur hearings, the House Committee on Science and Astronautics issued a report.[53] The committee emphasized Centaur's importance to the space program: the new vehicle was needed to "fill the performance gap between the Atlas-Agena and the Saturn class vehicles for space missions." It had been assigned important programs by both NASA and the Department of Defense. About twelve operational Centaurs a year had been planned to accomplish these missions. The committee noted that in addition to mission responsibilities, Centaur had an important role in research and development because "virtually all upper stages of large future space vehicles, both chemical and nuclear, are presently intended to utilize hydrogen as fuel." Centaur would provide essential information on the handling, storage, and firing of liquid hydrogen. "In short, Centaur is performing a technical pioneering function which is considered vital to the future of much of the Nation's space effort."[54] That effort included the piloted flight to the Moon as the third stage of the Saturn C-1. Without liquid-hydrogen upper stages, the weight of the hardware needed to accomplish the Apollo mission would double. In reviewing the history of the program, the report revealed that in the summer of 1960, NASA had enhanced the importance of Centaur to the space program when it funded other projects that used liquid hydrogen, such as the Rover nuclear rocket and the upper

---

[52] Ibid., 68.

[53] House Committee on Science and Astronautics, "Centaur Launch Vehicle Development Program," Report 1959, submitted by George P. Miller, 87th Congress, 2 July 1962.

[54] Centaur Report, 2 July 1962, 2.

stages of the Saturn and Nova vehicles. Given the program importance of Centaur, the committee demanded to know why it had not been assigned DX priority.

The committee placed much of the blame for Centaur's troubles on General Dynamics. It concluded that because of General Dynamics's preoccupation with the Atlas ICBM program, Centaur had not received the company's full attention. Until the reorganization under Grant Hansen, the company had resisted pressure from NASA to address managerial problems. The committee was especially critical of the company's failure to anticipate the problem with the liquid-hydrogen leakage in the intermediate bulkhead until after manufacture and shipment to the Cape. This "fundamental mistake in cryogenics engineering" should have been discovered much earlier through testing. The committee also criticized General Dynamics's carelessness in the location of test stands. A second Centaur had been damaged when an Atlas on a nearby test stand exploded.

But the committee's indictment was not limited to the contractor. It was also critical of Headquarters and Marshall for "weak and ineffective" management. They had allowed personnel responsible for the program to rely on troubleshooting as difficulties arose, rather than anticipating them. The delay of Centaur had compromised important missions like the Air Force's Advent Program. It was necessary to launch NASA's Mariner payload on the less powerful Agena. Surveyor, originally to weigh 2,500 pounds, had to be redesigned for a lower weight of 2,100 pounds, at a loss of $20 million to taxpayers.

What particularly galled the committee was the failure to reconcile the differences in design philosophy between Marshall and General Dynamics, especially as it related to the thickness of the intermediate bulkhead. The report stated:

> The subcommittee would not presume to decide the technical question of what consti-
> tutes a proper margin of safety in rocket construction, though the evidence seems to
> indicate that a more conservative policy might have saved both time and money in the
> Centaur development program. While it may not be possible to draw a final conclu-
> sion to the effect that one design philosophy is right and another is wrong, it is obvious
> that only one can prevail. The subcommittee considers it inconsistent for the major
> industrial contractor to adhere to one approach while the Government agency respon-
> sible for direction of the same development program adheres to another. Further, the
> subcommittee considers it important that industrial contractors follow agency specifi-
> cation, policies, and recommendations immediately upon becoming aware of them.[55]

The subcommittee recommended a thorough reevaluation of the program by NASA, including a resolution of the question of DX priority. Finally, it ordered an investigation of General Dynamics's billing by the General Accounting Office to find out whether "the interests of the

---

[55] Ibid., 13.

Centaur Assembly line at General Dynamics, 1962. (Courtesy of Lockheed Martin)

Government have been adequately protected under the contracts."[56] Of all NASA programs at that time, only Centaur and Surveyor had received this level of intense scrutiny and censure.

A Key Decision

Although the House Committee Report stopped short of recommending cancellation of the Centaur program, pressure was building within NASA to abandon the troubled liquid-hydrogen upper stage. Opposition to the Centaur program came chiefly from MSFC and JPL. Both Centers had earned reputations as preeminent authorities in questions related to rocket propulsion and spacecraft development. Their opinions carried weight with Congress and the American public. Von Braun and Brian Sparks, Deputy Director of JPL, were working behind the scenes to mount a campaign for the cancellation of Centaur in favor the Saturn C-1-Agena.

NASA Headquarters refused to give in to this pressure because of the high priority of Centaur within NASA. A memo from the Deputy Associate Administrator to Robert Seamans unequivocally stated that Centaur was needed to support Saturn development and carry out NASA's robotic planetary missions. The solution was not cancellation, but new management: "For the Centaur to

---

[56] Ibid., 12.

succeed, NASA must have a capable project management team to direct its efforts and to direct the contractor."[57] As a first step, NASA named Vincent Johnson Centaur program manager in June 1962. Johnson, a graduate of the University of Minnesota, had served as a physicist at the Naval Ordnance Laboratory during World War II. After a career in the Navy Bureau of Weapons, he came to NASA in 1960 as program manager for the Scout, Delta, and Centaur launch vehicles. In 1964, Johnson became director of the launch vehicle and propulsion programs in the Office of Space Science and Applications, where he remained until 1967. In Johnson, Centaur gained the undivided attention of a capable Headquarters executive with a firm grasp of engineering.

Johnson and Homer Newell immediately approached Abe Silverstein to see whether Lewis Research Center would be willing to take over the troubled program. Silverstein had played a signal role in shaping NASA under Glennan, but when James Webb took over as NASA Administrator in 1961, Silverstein's influence at Headquarters began to wane. After the decision to fund the giant Saturn V rocket for a piloted lunar landing, Webb wrestled with the decision of whom to choose to head the Apollo program. The two top candidates were Silverstein and von Braun. Webb passed over von Braun, reputedly because of the adamant opposition of his Deputy Administrator, Hugh Dryden. Put off by Silverstein's autocratic management style and his insistence on having complete control of the Apollo program, Webb asked Brainerd Holmes of the Radio Corporation of America to head the program. Silverstein accepted the post of Director of Lewis Research Center.[58]

Silverstein returned to Cleveland in November 1961, shortly before fellow Ohioan John Glenn's historic flight. During the nearly four years Silverstein had been away, Lewis Research Center had vacillated between whether to go after a significant piece of the space program or continue to focus on its forte—aircraft propulsion research. Many engineers feared that the quality and autonomy of their research might be compromised by the pressures of managing space missions, while others fumed that the Center would be left out of exciting developments in the new field of rocket propulsion. One engineer warned that JPL was aggressively expanding its rocket activities, particularly in cryogenics. "These programs," he wrote, "smack of the same concepts as have been pursued by [Lewis] with hydrogen-fluorine. JPL is also in a dither to get into development of electrical propulsion devices."[59]

Silverstein ended the debate through a sweeping reorganization. He divided the laboratory into two parts, making Bruce Lundin Associate Director for Development and Eugene Manganiello Associate Director for Research. For the next fifteen years, Lundin played a pivotal role in the management of Center space programs. Lundin had served as head of the Engine

[57] Thomas F. Dixon to Robert C. Seamans, Jr., 19 September 1962, Centaur file, NASA Historical Reference Collection.

[58] Henry Lambright, *Powering Apollo: James E. Webb of NASA* (Baltimore: The Johns Hopkins University Press, 1995), 108.

[59] Report by Howard Douglass, "The NASA Advanced Technology Programs for Liquid and Solid Propellant Rocket Engines,"15 January 1962, RG 255, NARA, Box 254.

Research Division in charge of testing full-scale turbojet and ramjet engines from 1952 to 1957. He was enthusiastically behind the effort to win a major piece of the space program. The reorganization was intended to keep research and development entirely separate. By March 1962, the development side of the laboratory was growing by leaps and bounds. Bruce Lundin set up a centralized office to handle technical support for four chemical rocket development programs: the M-1 engine for a proposed giant Nova rocket by Walter Dankhoff, the J-2 engine for Saturn by Ward Wilcox, the F-1 engine by Fred Wilcox, and the RL10 engine for Centaur by Eugene Baughman.[60]

In August 1962, Vince Johnson called Silverstein and Lundin to Headquarters to ask whether they were willing to take over management of the entire Centaur program. Lundin recalled that they were not particularly enthusiastic: "Abe said, 'Well someone's got to do it,' and I said, 'Well, we'll do our best.'"[61] The decision to move Centaur from Marshall proved controversial. There were technical, organizational, financial, and political problems. "Everybody at Lewis thought we were crazy to take that on because it was such a headache," said Bruce Lundin.[62] The task was monumental, and "everything that could be wrong was present."

Outsiders shared this opinion. An article in the *Washington Evening Star* viewed the transfer as an "action that smacked of desperation."[63] Journalist William Hines further wrote, "In its failure to meet deadlines, specifications and cost estimates, Centaur has poorly qualified for the title of Space Age Turkey No. 1." He thought it unlikely that even with a DX priority could Centaur be developed in time to prove useful for either the robotic space program or the program in planetary sciences. Because of Centaur program delays, NASA had to send the first Mariner probe on Agena. The reduction in the amount of weight the launch vehicle could carry made Mariner "an engineering triumph but a scientific starveling." Reflecting the "unspoken opinion" of many people in the aerospace community that Centaur would be "obsolete before it flies," the journalist favored using the Saturn C-1 for future scientific missions. He questioned whether another "stripped-down" Mariner would be worth flying to Mars. Bad press and skeptical Congressmen would continue to dog Centaur even after the first Surveyor touched down on the Moon.

Even after Silverstein had agreed to take over the program, von Braun and Sparks continued to urge cancellation of Centaur. In correspondence with Headquarters,[64] von Braun and Sparks

---

[60] Bruce Lundin to Associate Director, Lewis Research Center, "Centralization of the Center's Technical Support to Chemical Rocket Development Program," 7 March 1962, RG 255, NARA, Box 254. The J-2 and the M-1 engine for Nova both used liquid-hydrogen fuel. Nova was canceled after it was decided to use the less energy-intensive lunar orbit rendezvous (LOR) for Apollo.

[61] Interview with Bruce Lundin by Virginia Dawson, 7 March 2000.

[62] Ibid.

[63] William Hines, "Centaur or Saturn–Which?" *Washington Evening Star* (15 October 1962): 10.

[64] Brian Sparks to Homer E. Newell, 21 September 1962, NASA Historical Reference Collection.

asked "clearly, strongly, and unequivocally" that the Saturn C-1-Agena be substituted for Atlas-Centaur in carrying out unpiloted lunar and planetary missions. Sparks contended that Centaur had created a "deplorable situation" for both the lunar and planetary programs. NASA needed the Saturn C-1-Agena to counter the "doggedly determined effort of the Soviets" and restore national pride and international prestige. In a detailed memo appended to his letter, Sparks compared Centaur to C-1-Agena. Despite the fact that Centaur development was in full swing, while Saturn C-1 was only in the design stage, Sparks thought that Saturn C-1 personnel would be highly motivated by their association with the piloted flight program. In contrast, the fear of cancellation of Centaur would demoralize the Centaur team and slow their efforts.

Newell patiently responded that projections for completing the development of Saturn C-1-Agena were overly optimistic. It had been considered and rejected. He agreed that Centaur development had been "fraught with difficulties," but he expected that the transfer to Lewis would "breathe new life into it, and that it will be able to fill the vital needs of our lunar and planetary program on a timely basis."[65] He thought it unlikely that the untested S-4 stage could be adapted to the S-1 stage and integrated with Agena before Centaur could be made flightworthy. During this debate, the transfer to Lewis Research Center in October 1962 went quietly forward.

If the decision to transfer the program represented an ultimatum for General Dynamics, it was an equally important turning point for Headquarters. In his detailed analysis of the Surveyor and Lunar Orbiter programs in 1972, management consultant Erasmus Kloman provided some useful insights into the change in Centaur management. He speculated that by refusing to cancel Centaur, Headquarters (heavily staffed by former Lewis Research Center engineers at this time) asserted its prerogative to determine launch vehicle policy. He wrote:

> Headquarters, after carefully reviewing the situation, confirmed its position that the Centaur concept was both technically feasible and essential to the launch vehicle program for the space effort. It thus rejected the recommendation of senior management at MSFC and JPL. Responsibility for Centaur was transferred abruptly then to Lewis Research Center. This was interpreted as a rebuke to MSFC and a signal to the other Centers that they could not back out of major development commitments assigned by Headquarters.[66]

An even more convincing reason for keeping Centaur in NASA's stable of rockets was its role in the Apollo program—both as the upper stage for Surveyor and as a means to test liquid

---

[65] Homer E. Newell to Brian Sparks, 23 October 1962, NASA Historical Reference Collection.

[66] Erasmus Kloman, *Unmanned Space Project Management: Surveyor and Lunar Orbiter* (Washington, DC: NASA SP-4901, 1972), 32. See also an unpublished draft, NASA Marshall Space Flight Center archives, chapter 2, 62–64.

hydrogen's feasibility in the Saturn liquid-hydrogen/liquid-oxygen upper stages. Only its association with the Apollo program kept the Centaur program alive.

President Kennedy made this clear to NASA Administrator James Webb several times during a meeting in November 1962 about supplemental appropriations. The meeting was attended by Webb, top NASA staff, and Jerome Wiesner, Special Assistant to the President. At several points in this meeting, Kennedy emphasized that beating the Russians to the Moon was the only reason that he was willing to contemplate the enormous appropriations requested by NASA. Space program contributions to science were a distinctly lower priority, for which reasonable sums of money could be spent. "But we're talking about these *fantastic* expenditures which wreck our budget," he said, which could only be justified by the goal of overtaking the Russians, "and demonstrate that starting behind, as we did by a couple of years, by God, we passed them."[67]

When Kennedy asked Webb what the nation could expect from Centaur, he responded that Centaur would save 50 percent of the cost of the planetary shots, presumably a reference to the high cost of developing the Agena D, the upper stage favored by von Braun. "This was worth fighting for,"[68] he said. But knowing that Kennedy would not accept this as the sole justification for keeping Centaur in the rocket lineup, Webb immediately added that Centaur would allow the government to work out problems on a small vehicle before testing the far larger liquid-hydrogen Douglas upper stages for Saturn.[69] Webb did not excuse the poor job of management of the Centaur program by Marshall or General Dynamics's failure to recognize the magnitude of the problems associated with liquid hydrogen. However, he was obviously furious that von Braun had taken the unprecedented step of appealing by letter directly to Congress for the cancellation of Centaur in favor of the Saturn C-1-Agena.[70] Other participants in the meeting emphasized the absolute necessity for Surveyor as the justification for the Centaur program.

One of the questions that Surveyor was charged with solving was the composition of the Moon's surface. Cornell scientist Thomas Gold had argued that billions of years of bombardment by asteroids had pulverized the Moon's surface into a fine powder about thirty feet deep. He speculated that the low bearing strength of the surface would cause the Apollo lander to founder.[71] Others questioned this thesis, but lack of actual data increased the risk of sending astronauts to the Moon. To emphasize the connection between Surveyor and the Apollo program, Jerome Wiesner told President Kennedy,

---

[67] Transcript of Presidential Meeting in the Cabinet Room of the White House, 21 November 1962, John F. Kennedy Library, President's Office files, tape no. 63, NASA Historical Reference Collection, page 31 of transcript.

[68] Transcript, 19.

[69] Ibid.

[70] Ibid, 18. This letter has not been found.

[71] Robert S. Kraemer, *Beyond the Moon: A Golden Age of Planetary Exploration 1971–1978* (Washington: Smithsonian Institution Press, 2000), 29.

We don't know a damn thing about the surface of the Moon. And we're making the wildest guesses about how we're going to land on the Moon and we could get a terrible disaster from putting something down on the surface of the Moon that's very different than we think it is. And the scientific programs that find us that information have to have the highest priority. But they are associated with the lunar program. The scientific programs that aren't associated with the lunar program can have any priority we please to give 'em.[72]

Despite the problems associated with the development of Centaur, the need to solve the question of the composition of the lunar surface gave Centaur the priority it needed to avoid cancellation. Making Centaur an inextricable part of the lunar landing effort was a calculated risk on the part of NASA Headquarters. Webb refused to yield to pressure exerted by Huntsville and JPL for cancellation. Now he was depending on Abe Silverstein, with all the resources of a research laboratory at his disposal, to prove that it was the right decision. Silverstein's stubborn advocacy of liquid hydrogen would now be put to the test.

---

[72] Transcript, 29.

Chapter 3

"Abe's Baby"

"These guys had a capacity, an ability, an energy that is just not describable . . . . They gave up their lives, their families, they gave up everything. They had total commitment. It was an era, a period of time, that was absolute, total commitment."

–Harlan Simon, Lewis Research Center

At Headquarters, Abe Silverstein had experienced the risks and challenges of managing large-scale projects. He seemed to have no doubt that the NACA-trained engineers at Lewis could learn how to monitor General Dynamics and deal with the prickly payload specialists at JPL–especially since Lewis already had a unique expertise in liquid-hydrogen technology and a Director who would let nothing stand in the way of Centaur development. A down-to-earth midwesterner who did not suffer fools gladly, Silverstein was known for his keen engineering insight. In contrast with the tepid efforts at Marshall, NASA under Silverstein would move heaven and Earth to turn Centaur into a reliable launch vehicle. Harlan Simon, a young procurement officer, called Centaur "Abe's Baby," pointing out that both dedication and fear of Silverstein's famous temper drove the program: "You just didn't dare let him down."[1]

Management of Centaur would prove both a trial and a triumph for Lewis Research Center. Silverstein announced that he would personally direct the project with J. Cary Nettles as acting Centaur project manager. This was a temporary arrangement until permanent leadership could be established. Initially, forty one Lewis personnel were assigned to the project. Nettles, a graduate of Louisiana State University with a degree in electrical engineering, had been among the contingent of NACA engineers transferred from Langley to Lewis in 1941. As chief of the Flight Projects Branch, he was already serving as the Lewis representative on a NASA project to study the effects of zero gravity on liquid hydrogen at Wallops Island, Virginia. Nettles immediately uncovered a chaotic contracting situation. A large number of changes, not only in the contract with General

---

[1] Virginia P. Dawson, *Engines and Innovation: Lewis Laboratory and American Propulsion Technology* (Washington, DC: NASA SP-4306, 1991), 192.

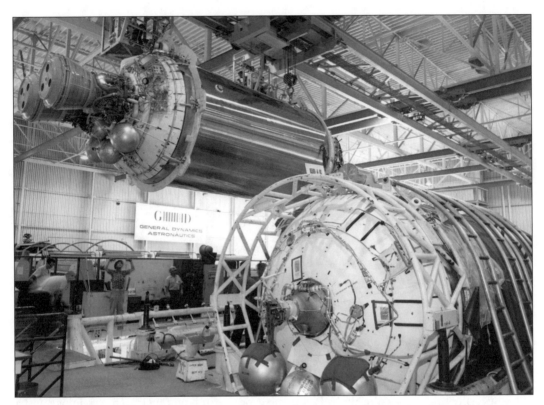

AC-2, suspended by an overhead crane, is readied at the General Dynamics factory for shipment to Cape Canaveral in 1964.
(Courtesy of Lockheed Martin)

Dynamics, but also with Minneapolis-Honeywell for the guidance, had driven up the cost of the program.[2] One of the priorities was to get the escalating costs of the program under control.

A terse telegram from NASA Headquarters to Silverstein announced that the transfer of Centaur from MSFC to Lewis would occur between 8 October 1962 and 1 January 1963. About forty Marshall engineers assisted in the changeover. They were not inclined to linger in Cleveland; they hastened back to the milder climate of Huntsville before the first snows flew. As Centaur documentation began to accumulate, a skeleton project office was set up in the hangar.

NASA set an ambitious target date of late 1964 for the first Surveyor mission. Space Technology Laboratories (STL) agreed to conduct an independent guidance analysis. This was apparently at the insistence of JPL, which did not trust the competence of either General Dynamics or Lewis in this area.[3] Many of the engineers at STL, a spinoff of the Ramo-Wooldridge Company, had served as technical monitors for Atlas. Silverstein pressed NASA to transfer management of the RL10 engine to Lewis at this time. Von Braun refused, presumably

---

[2] Communication from J. Cary Nettles to Virginia Dawson, June 1999.

[3] Interview with Richard Martin, 5 June 2000.

because of the RL10 engine's role in the development of the S-IV liquid-hydrogen upper stages for the Saturn C-1. Whether Atlas procurement would continue to be managed by the Air Force as government-furnished equipment also proved contentious.[4] Silverstein placed Ed Baehr in charge of an Atlas office at Lewis.

Under Silverstein, for the first time, the government would take a strong role in managing Centaur. Silverstein announced that the Research & Development phase for Centaur would use eight test vehicles. This phase was expected to last through 1964 (it was not actually completed until October 1966). Only after they had proven the reliability of Centaur would Silverstein permit a series of eight Surveyor spacecraft to be launched. To reduce technical complexity and interface problems, he decided that the first seven Surveyors would carry essentially the same payload with different experiment packages. This decision meant that many of the scientific experiments planned for Surveyor had to be sacrificed.

Three months later, in December 1962, Lewis Research Center took over Agena, another troubled launch vehicle program previously managed by Marshall. The transfer was precipitated by the failure of the fifth Ranger spacecraft to reach the Moon. Headed by Seymour Himmel at Lewis, the Agena program never received the same attention as Centaur, but its management contributed to the growing launch vehicle expertise among NASA staff. Not only did the program launch Atlas-Agenas from the Eastern Test Range, but it also launched Thor-Agenas from Vandenberg Air Force Base in California for polar and Sun-synchronous orbits. Atlas-Agena missions included the first closeup pictures of the Moon, Mars, Venus, early communications satellites such as Echo, and large weather satellites like Nimbus.

Atlas-Agena B could place about 5,000 pounds in Earth orbit and lift 750 pounds to escape velocity. Atlas-Agena B launched the last series of Ranger spacecraft, which succeeded in sending back medium- and high-resolution photographs of the Moon's surface. Between 1964 and 1968, Atlas-Agena D launched twenty payloads for NASA and the Department of Defense, keeping the country's lunar and planetary science program alive while the nation waited for the more powerful Centaur.[5]

In February 1963, the General Accounting Office (GAO) completed its investigation of the Centaur program. The report, based on an internal NASA Headquarters investigation by William Fleming, was extremely critical. Fleming had worked at Lewis prior to accepting a post at Headquarters under Silverstein. His report reflected criticism of General Dynamics that came directly from the experience of various NASA staff, including what they perceived as the contractor's significant lapses leading up to the failure of F-1. The Fleming report concluded that both NASA and General Dynamics should have predicted the rupture of the

[4] Thomas Dixon to Robert Seamans, Jr., 4 October 1962, John Sloop papers, NASA Historical Reference Collection.

[5] *Lewis News* (18 August 1967, 15 October 1965, 24 November 1967). NASA launched thirty-eight Agenas between 1961 and 1967.

weather shield from available test data. The GAO's report called this "not merely an indica-tion of error or failure to anticipate a difficulty but rather a critical dereliction in management, since the problem was perceived, tests undertaken to resolve the problem, and a report on test results [by Ames Research Center] furnished but not completely utilized."[6] GAO qualified this sharp criticism by noting that on average, twenty flights were usually needed to establish a 50-percent reliability for a new launch vehicle. Less significant technical lapses, such as the failure of General Dynamics to provide an independent flight abort system for both Atlas and Centaur, were also noted. Additionally, the GAO accused General Dynamics of improper billing practices, including failure to provide adequate documentation for fees.

The GAO was equally critical of what it called Marshall's "sloppy" method of contracting. Marshall had allowed General Dynamics to set unrealistic launch dates and had failed to monitor how much weight Centaur could lift. The GAO concluded, "The most penetrating business crit-icism related to the substantial number of uncompleted contractual actions and generally poor documentation which, though originally attributable to poor AF [Air Force] management, was perpetuated under Marshall management and is only now being remedied by Lewis."[7]

Building Confidence in Centaur

Silverstein insisted on a system-by-system review to help bring Lewis engineers up to speed and allow problems to surface. They found problems in every area. Their first task was to build an effective working relationship with General Dynamics engineers. In February 1963, Silverstein appointed David Gabriel manager of the program. The project was first located in the basement of the administration building, where Silverstein could check on it frequently. Under Gabriel, he assigned three assistant managers: Edmund Jonash, in charge of test engineering, structures, and vehicle propulsion; Cary Nettles, oversight of guidance, reliability, quality assurance, and ground support equipment; and Ronald Rovenger, head of field operations. Rovenger, one of the few Marshall staffers to transfer with Centaur, headed the NASA field office at General Dynamics. Under Lewis management, his office increased to forty NASA engineers. Gabriel and Jonash, like Nettles, were old NACA hands. Gabriel, a native of Lakewood, Ohio, had graduated from the University of Akron in 1943. He was already familiar with Centaur problems through zero-gravity test work. At this time, he headed the Advanced Development and Evaluation Division, which was providing technical support for a joint Atomic Energy Commission-NASA nuclear rocket program called Project

[6] Memo from Nathaniel H. Karol to Thomas W. Bugher, Acting Chief, Contract Management Branch, "GAO Exit Conference at NASA Headquarters, 7 February 1963." Box 9, NASA Glenn History Office Archives.

[7] Ibid.

Meeting of the Centaur team at General Dynamics, early 1964. Standing, left to right: unknown; Dr. Drew Katalinsky, Director of Engineering for Centaur; Grant Hansen, General Dynamics Vice-president and Centaur project head. Seated, left to right: Ed Jonash, Dave Gabriel, and Cary Nettles, leaders of the team from Lewis Research Center; Major Joe Heatherly, U.S. Air Force (USAF); and Ron Rovenger, head of the NASA Resident Office at General Dynamics. (Courtesy of Lockheed Martin)

Rover.[8] Jonash, a native of Kenton, Ohio, with a degree from MIT in chemical engineering, had participated in testing the RL10 engine in Lewis's Propulsion Systems Laboratory. Because the group spent much of the week in San Diego, staff meetings were held on Saturdays, the only time most people could be expected to be in Cleveland.

Lewis staff for the project increased to 150 people. By 1964, Centaur had a budget of $10 million for in-house support. Silverstein assigned more than forty engineers and support staff to monitor the General Dynamics plant in San Diego and at Minneapolis-Honeywell. Under the new management, Centaur became a major program for General Dynamics, approaching 50 percent of total effort in the Astronautics Division.[9]

---

[8] David Gabriel became Centaur Project manager in 1963. In 1965, he left NASA to take a job for Bell Aerosystems. In 1967, he became Deputy Manager of the Atomic Energy Commission (AEC)-NASA Office Space Nuclear Propulsion Office (SNPO).

[9] *1966 NASA Authorization,* Hearings before the Subcommittee on Space Science and Applications, H.R. 3730, pt. 3, March 1965, 1145.

A contingent of young NASA scientists and engineers recruited to work on the space program reveled in the opportunity to learn in an environment that pushed them intellectually and gave them responsibility early in their careers. This infusion of raw talent transformed Lewis Research Center. Lawrence Ross, who became Director of the Center in the 1990s, began as a design and test engineer for Centaur. He investigated Centaur from stem to stern, learning how to rewire it, design test equipment, and run tests. This early hands-on experience provided a unique foundation for his NASA career.[10] Work on the sloshing of propellants brought Andrew Stofan, another future Director, into the Launch Vehicles Division. Ross, Stofan, and other recent graduates were trained and mentored by men who had NACA research backgrounds, as well as extraordinary technical insight, stamina, and blind faith that Centaur would work. Harlan Simon, a thirty-two-year-old lawyer at the time he was hired in the Centaur Procurement Office, marveled at the dedication of Gabriel, Jonash, and William (Russ) Dunbar, who were then in their forties. "These guys had a capacity, an ability, an energy that is just not describable," he recalled. "They gave up their lives, their families, they gave up everything. They had total commitment. It was an era, a period of time, that was absolute, total commitment."[11]

In December 1962, Centaur received DX priority because of its connection with the Apollo program. This made keeping contractors to strict deadlines much easier, since the program had national priority in terms of supplies and personnel. The most pressing concern was the need to untangle the complex contractual relationship between the government and General Dynamics. NASA inherited Centaur cost-plus-fixed-fee contracts from the Air Force. This included six Air Force contracts and one NASA contract; the latter contained 120 changes with 47 in progress at the time of the transfer. The process of making necessary NASA-directed changes needed to be addressed.[12] Each design change required the approval of the Air Force System Project Office. The Air Force resisted changes to Centaur because they often had an impact on the Atlas design, which the Air Force considered "operational." The Air Force complicated relations between General Dynamics and Lewis, creating bottlenecks in Centaur development and causing friction between the two government agencies. Ultimately, when the Air Force Centaur contracts were turned over to Lewis, NASA was allowed to purchase a "bare bones" Atlas from the Air Force so that the Air Force could still control Atlas manufacturing. Atlas could then be modified to support Centaur requirements.

---

[10] Interview with Larry Ross by Mark Bowles, 29 March 2000.

[11] Interview with Harlan Simon by Virginia Dawson, 20 March 1985.

[12] Thomas F. Dixon, Deputy Associate Administrator to Robert Seamans, Associate Administrator, 4 October 1962, files of Cary Nettles. The GD Centaur contracts had about $98 million in changes, most of which occurred prior to the contracts' transfer to Lewis. Eighty-three percent of the changes were attributed to directed changes and about 17 percent classed as overruns. See *1966 NASA Authorization*, Hearings, March 1965, 1145.

Because Lewis Research Center had never managed large contracts with industry, it had few procurement officers. The Center quickly hired about half a dozen lawyers from the Cleveland Ordinance District, Department of the Army. Nettles insisted that Len Perry, a talented African American graduate of Cleveland Marshall Law School, handle Centaur contract negotiations with General Dynamics. Perry worked under Edward Hicks, chief of procurement for Lewis. The procurement team included Harlan Simon, Richard Campbell, William Monzel, Marvin Clayton, Richard Proctor, William Brahms, and Robert Hill. Because Silverstein insisted that a new contract would be necessary, Perry and his counterpart at General Dynamics, T. C. Courington, began at once to review the contracts.

Silverstein personally conducted Centaur negotiations. He insisted that no Centaur would be flight-tested until "the probability of success approaches that necessary for manned space flight."[13] Bruce Lundin recalled that General Dynamics, accustomed to Air Force weapons procurement, at first objected to being held to a higher standard. "I said, 'Naturally that's impossible, but you won't get near it unless you require it.' This is space flight and that requires perfection at every point."[14]

The new contract, hammered out over a period of two and a half years, became NAS3-3232, or "32 squared," as NASA insiders called it. Jim Dempsey arrived in Cleveland to personally meet with Silverstein as negotiations drew to a close. They ended with a walk in the garden of Guerin House, a former private home converted to a center for meetings at Lewis. The two men agreed on the company's fixed fee, or profit of $31,293,524. The new contract, signed in April 1964, combined the hodgepodge of contracts inherited from the Air Force and Marshall into one cost-plus-fixed-fee contract for fourteen vehicles and twenty-one test articles. It included design, fabrication, and modification of ground-support equipment for the General Dynamics factory, the Atlantic Missile Range Complex 36A, the Point Loma Test Site, the Sycamore Canyon Test Site, the rocket site at Edwards Air Force Base, and NASA Lewis test sites at Plum Brook and its space power facility. The estimated cost of the contract was $321,058,005, plus the fixed fee. This contract covered known items, a specific number of launch vehicles, and equipment. In addition, Silverstein insisted on a second contract to permit greater flexibility in dealing with unforeseen problems that were likely to crop up during the development of new technology under the pressure of a tight deadline. This Sustaining Engineering and Maintenance (SE&M) contract covered modifications and improvements and eliminated the need for costly change orders. The SE&M contract covered the costs of the work of about two hundred General Dynamics engineers. NASA paid the

---

[13] "New NASA Management Concept Aims for Higher Centaur Reliability," *Aviation Week & Space Technology* (21 January 1963): 37.

[14] Interview with Bruce Lundin by Virginia Dawson, 7 March 2000.

direct costs of "technical directions" from Lewis to General Dynamics, plus an incentive fee that was determined unilaterally by Lewis. Silverstein himself made the quarterly evaluation of the contractor's performance. When Bruce Lundin became director of Lewis in 1969, he continued the practice.[15]

During the time it took to get the contracts straightened out, Lewis and General Dynamics engineers continued to drive the technical aspects of the program forward. Dave Gabriel used PERT to force the pace of development. PERT also found favor with Grant Hansen. In an article in *Aerospace Management* in 1963, Hansen noted that PERT had made determining the costs of the program's various subdivisions more realistic.[16] Hansen liked the attitude of the new people in charge at NASA. He recalled that when contract monitors from Marshall had come out to San Diego, they had made it clear that they were there to check up on them, with the implication that they could not be trusted to do the job right. Lewis management simply said, "We're here to help you do it."[17] General Dynamics engineer Denny Huber noted that when the Lewis people took over Centaur, they "wanted to operate as a team with [General Dynamics] and not act as Lord and Master over us; what a breath of fresh air."[18] By degrees, Lewis engineers and their counterparts at General Dynamics learned to respect each other. As General Dynamics launch conductor Roger Lynch put it, "What they did better than the Marshall guys, is that they started to identify people they could trust."[19]

Direct Ascent

Silverstein's key technical decision was the abandonment of the requirement for a two-burn mission for Surveyor. He insisted that first the feasibility of a single-burn mission—a straight shot to the Moon—had to be proven. Only after direct ascent was successful would he agree to attempt a two-burn mission with a coast period and restart in space. He reasoned that with less technical complexity, there was less chance of failure. Direct ascent would save weight because it did not require extra fuel and equipment to settle the propellants during the coast and for attitude control.

The previous summer, while he was considering whether Lewis Research Center should take over the program, Silverstein had sent Art Zimmerman on a reconnaissance trip to California to discuss whether direct ascent was feasible. Zimmerman had contacted George Solomon, the head of trajectory analysis at Space Technology Laboratories in Redondo Beach. He called on Victor Clark, a

---

[15] Interview with Bruce Lundin. These contractor performance reports would be revealing if they could be found. They do not appear to be part of the NASA Glenn records. Congressman Karth did not think an SE&M contract for Centaur was appropriate, since $35 million for SE&M was actually for development costs. See *1967 NASA Authorization*, Part 3, 583.

[16] Philip Geddes, "Centaur: How It Was Put Back on Track," *Aerospace Management* (April 1964): 24–29.

[17] Interview with Grant Hansen by Virginia Dawson, 6 June 2001.

[18] Communication to the authors from Denny Huber, 2 March 2002.

[19] Interview with Roger Lynch by Virginia Dawson, 5 June 2000.

scientist at JPL, and Frank Anthony, assistant chief engineer of flight mechanics at General Dynamics. None of the three were entirely convinced; it was fortuitous that the Moon's orbit happened to be in a place within its eighteen-year cycle that made direct ascent possible in the mid-1960s.[20]

The decision by Silverstein to abandon the two-burn requirement initially shocked the rank and file at General Dynamics, a contractor accustomed to the relatively hands-off management style of the Air Force. At least two burns had previously been needed for both the Advent mission and Surveyor. The company had focused on the more difficult approach. Now Silverstein told General Dynamics that the company had only one requirement on Centaur—to show that the system worked. Deane Davis recalled that people at General Dynamics reacted as though they had been asked to give up one of the Ten Commandments. "It didn't make any difference whether we were receptive or not," he said. "Abe had made up his mind."[21]

This change to direct ascent also created consternation among scientists at JPL because it reduced the number of launch windows, or available dates for launch. A January 1963 memo from Benjamin Milwitzky, chief of the Surveyor program, to Oran Nicks, director of NASA's Lunar and Planetary programs, strongly recommended that the two-burn option be emphasized in the development phase of the program. He worried that if the Centaur program were delayed until the period between October 1965 and February 1966, Surveyor would have to land in the dark. "If these Surveyors have to land in the dark and survive until lunar daylight before beginning operations on the Moon, the probability of completing the scientific mission will, most likely, be significantly reduced. Also, the possibility of obtaining TV pictures on the way down will be eliminated," he wrote.[22] Milwitzky told Nicks that JPL (probably reflecting reservations of Marshall engineers) also questioned the reliability of the Centaur guidance system because the Minneapolis-Honeywell hardware had not yet been tested under environmental conditions. Nevertheless, the tone of the memo was positive. These problems could be solved.[23]

Although people at JPL and General Dynamics found Silverstein's management style to be heavy-handed, they soon discovered its benefits. Marshall and JPL had wrangled over the change of weight specifications for Surveyor made necessary by the continuing problems associated with the development of Centaur. The first order of business was to fix the weight specifications once and for all. Silverstein asked Hugh Henneberry to head a Performance Trajectory Group (PTG) made up of people from both Lewis and General Dynamics. Their charge was to make sure that the weight requirements were carefully negotiated and mutually acceptable. Other

---

[20] Interview with Art Zimmerman by Virginia Dawson and Mark Bowles, 5 August 1999.

[21] John Sloop interview with Aerospace Division, Convair/General Dynamics, 29 April 1974, NASA Historical Reference Collection.

[22] Memo, Benjamin Milwitzky to Director, Lunar & Planetary Programs, 21 January 1963, Surveyor files, NASA Historical Reference Collection.

[23] Ibid.

members from Lewis included Zimmerman, Jack Lee, Harold Valentine, Carl Wentworth, Jack Brun, and Richard Flage. Jan Andrews, Frank Anthony, Robert Foushee, Joseph Garside, F. W. Koester, A. Rosin, and Richard Wentink participated for General Dynamics. Although Lewis took responsibility for all assumptions and ground rules, the effort was clearly based on the greater experience and expertise of General Dynamics. Out of this early government-industry collaboration came the report "Estimate of Atlas-Centaur Performance Capabilities for the Surveyor Mission," with trajectory calculations and performance ground rules for both direct ascent and parking-orbit ascent.[24] The introduction to the document stated:

> The philosophy adhered to throughout has been to provide a consistent set of rules defining a vehicle as it is expected to exist in 1965 when the first operational launch is anticipated. In many areas, insufficient data are available at present to allow a rigid definition of vehicle parameters. In spite of this, estimates have been made in all areas so that a complete configuration is defined. This, together with the assumptions and methods of the analysis can then form the basis for future changes or re-evaluations as they become necessary.[25]

Based on weight and other assumptions agreed to by the group, nominal payload capabilities of 2,257 pounds for a two-burn mission and 2,317 pounds for direct ascent were calculated for Surveyor missions. The Centaur vehicle for the two-burn mission had to be heavier because of the need to carry more fuel and equipment. The group suggested that improvements in a number of areas might allow the payload weights to increase by about 150 pounds. These weights were encouraging. In fact, when the missions were flown, the Surveyor model weighed 2,070 pounds, with the heaviest Surveyor carrying 2,295 pounds.

This preliminary study represented an important first step toward preparing Centaur for launch and gave Surveyor payload specialists at JPL definite figures on which to base spacecraft development. It was the first in a series of memoranda, issued monthly by General Dynamics, that included all changes to the vehicle up to the time of publication. Changes that had occurred the previous month were printed in red. Since the monthly memorandum always had a centaur on the front cover, the team began referring to the document as "The Horse" (a reference to the mythical Centaur—a horse with a man's upper torso and head). Roy Roberts, General Dynamics's principal guidance engineer on Centaur at that time, recalled that "The Horse" was a "very, very critical document because it documented for a large number of people the environments and the trajectory that we were flying."[26] It was distributed widely to the allied

---

[24] "Estimate of Atlas-Centaur Performance Capabilities for the Surveyor Mission," 11 February 1963, Kennedy Space Center ELV Resource Library. (The authors thank Art Zimmerman for lending his copy of this report.)

[25] Ibid.

[26] Interview with Roy Roberts by Virginia Dawson, 21 March 2001.

disciplines, such as thermal and loads. Under Marshall management, Centaur contractors and NASA managers had lacked this type of cumulative up-to-date documentation. "The Horse" reflected Centaur's continuing evolution and kept all participants informed and moving forward as a team. Both General Dynamics and Lewis engineers could run independent simulations. "It was a team relationship without ever being adversarial," Roberts remarked. "We were all in it together and doing the same thing."[27] The effort to approach problems in a collaborative spirit is what distinguished the program under Lewis management from that of Marshall. What Lewis engineers lacked in training and experience in project management they made up for in organizational and communication skills.

Although staff at the Jet Propulsion Laboratory had once disparaged NACA-trained engineers as myopically focused on "pushing known principles to the next decimal point," relations between members of the former research laboratory and employees of JPL became more cordial as Lewis engineers established their competence and won credibility.[28] Surveyor's managers at JPL found the weight and performance summary called "The Horse" especially helpful. Bill O'Neil, a young scientist assigned to the Surveyor trajectory and performance group at JPL headed by Elliot Cutting, recalled that not only did "The Horse" prove to be "a model document that really spelled out all the ground rules for everything,"[29] but it also influenced how JPL managed Hughes Aircraft, the contractor building Surveyor. JPL asked Hughes to produce a companion document modeled on "The Horse" that spelled out the spacecraft performance side. O'Neil said, "In fact, some people called it the pony. And other people called it the rider, because it was riding on it [Centaur]."

JPL and Lewis established a working group with responsibility for the interface between the launch vehicle and the payload. The group was charged with resolving problems related to interface design, interface testing, and system operations.[30] Once JPL and Lewis reached agreement on weights, launch times, and mission goals, they provided them to General Dynamics, where they could be calculated in greater detail. Zimmerman recalled, "With our optimization capability, we could design the trajectory and hand it over to them and show them this is the ideal way to fly the mission. Then they would take their huge program and tailor it to that flight mission. It usually came within 8 or 10 pounds of what we proposed."[31]

---

[27] Ibid.

[28] Clayton R. Koppes, *The JPL and the American Space Program* (New Haven: Yale University Press, 1982), 97.

[29] Interview with William O'Neil by Virginia Dawson and Mark Bowles, 9 June 2000.

[30] Technical Report 32-1265, Surveyor Project Final Report, quoted by Erasmus Kloman, typescript, 1971, Centaur chapter, 75, MSFC archives.

[31] Interview with Art Zimmerman by Virginia Dawson and Mark Bowles, 5 August 1999.

Centaur is offloaded from a C-133 transport plane at Lewis Research Center, June 1963. (NASA C-65194)

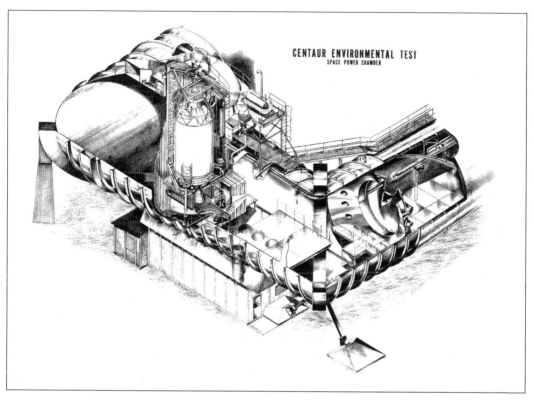

Drawing of the Centaur rocket placed in the Lewis Space Power Chamber, formerly the Altitude Wind Tunnel. (NASA Drawing 7709-EL)

"Abe's Commandment"

Silverstein's emphasis on ground testing represented a shift from the relatively hands-off approach of the Air Force. Greater supervision by NASA initially produced friction, especially since engineers at General Dynamics were far more experienced in rocket development than their Lewis counterparts. Grant Hansen, head of Centaur for General Dynamics, questioned his new Lewis managers' "willingness to test until everything was absolutely locked down, without any particular regard for an ultimate end date that had to be met." Deane Davis called testing "Abe's commandment." In order to enable General Dynamics to check the interface between Atlas and Centaur before shipment to the Cape, NASA now agreed to the construction of a ten-story, $6.8-million facility adjacent to the Kearny Mesa Industrial Park, despite the fact that General Dynamics already had an impressive array of test facilities.

Silverstein insisted that Centaur also be tested at Lewis. Testing began with components, followed by engine tests, then tests of the entire system, and, finally, full static tests of the entire flight configuration. Separation tests, shake tests, structural tests, nose-fairing tests, and insulation-panel tests all contributed to growing confidence that the Center had the facilities and expertise to monitor the contractor.[32] In Silverstein's view, even if there were only a one-in-a-thousand chance of failure of one component, it was better to test it than to risk failure. His technical perfectionism was legendary. To Silverstein, space vehicles required "not statistical accuracy but one hundred percent accuracy."[33] This could only be achieved by "extreme diligence and the concept that every piece of equipment that is taken aboard, every component, must be proven– environmentally checked so that it can live in the environment of space, the total environment of space, the vacuum of space, the temperatures of space." Only after systematic ground testing was completed was he willing to allow Atlas-Centaur to be launched.

Testing at Lewis not only provided independent verification of contractor performance, but also contributed new solutions to problems. The loss of F-1 had occurred within a tenth of a second after hydrogen gas had been vented from an opening at the top of the vehicle. A test program in the 8-by-6-foot supersonic wind tunnel at Lewis revealed that the hydrogen-venting system posed a fire hazard during flight. The solution was to design a vent fin or snout on the nose fairing that extended about 50 inches from the tank, just far enough away to keep the hydrogen gas from igniting along the hot surface of the vehicle.

As General Dynamics people began to see the concrete results of these test programs, respect began to replace the adversarial relationship between General Dynamics and the government.[34] Testing applied to whole systems, as well as individual components. The laboratory hastily converted the Altitude Wind Tunnel into a vacuum chamber in order to test a full-scale operational Centaur vehicle under environmental conditions to simulate spaceflight. Test engineers "soaked" a Centaur in the vacuum chamber to test the separation system by firing the retrorockets in the normal and failure modes. Lewis wind tunnels were also used to test the RL10 engine, while other liquid-hydrogen tests were carried out in the unique Rocket Engine Test Facility at the Center. Although needed for no more than 3 minutes, the flight gyros received 1,000 hours of testing, with the rationale that if the gyros worked for 1,000 hours, they would run several minutes more in space without a problem. Explosive bolts, which could not be tested, were carefully inspected for defects. If, as Lewis engineers often pointed out, there were 1,000 single failure modes on Atlas-Centaur, they were confident that many "bugs" could be uncovered with adequate testing.[35]

Silverstein believed that in addition to providing industry with test facilities, it was important for NASA to build extensive new facilities for rocket tests. Centaur presented special challenges because its engines had to fire in space. With the Centaur carrying one-of-a-kind multimillion-dollar payloads, how could engineers be sure that the Centaur engines would start in a

[32] Philip Geddes, "Centaur: How It Was Put Back on Track," *Aerospace Management* (April, 1964): 24–29.

[33] John Sloop interview with Abe Silverstein, 19 May 1974, NASA Historical Reference Collection.

[34] Personal communication to the author from J. Cary Nettles, 17 June 1999. See also Eugene Kloman, "Centaur," typescript, MFSC archives, 69–70.

[35] Interview with Bruce Lundin by Virginia Dawson, 7 March 2000.

near-vacuum while being subjected at different times to extremely hot and cold temperatures? To answer these questions, NASA appropriated funds for the construction of a unique facility called the Space Propulsion Research Facility, or "B-2," at the Plum Brook Station, located 55 miles from Lewis Research Center in Sandusky, Ohio. At the Plum Brook's "E Site," the Dynamics Research Test Center, an entire Centaur test vehicle could be mated to the Atlas, along with a test model of the Surveyor spacecraft. Important "bending mode tests" performed by Ted Gerus, head of Dynamics, and Robert P. Miller, project lead engineer, involved tanking the vehicle with simulated propellants and shaking it horizontally and vertically to test its structural integrity. The Atlas also received axial load tests at 248,000 pounds to simulate the forces on its thin skin during liftoff. It passed with flying colors.

These tests of the steel balloon tank structure provide another example of the cooperation between General Dynamics and Lewis engineers. David Peery, former head of the Aeronautical Engineering Department at Penn State University and author of an important textbook on aircraft structures, had developed a theory for structural strength available after the onset of local skin buckling. At General Dynamics, he showed by analysis that the Atlas-Centaur tank design— thought by many to be too radical—was actually unnecessarily conservative. According to his theory, it could withstand vastly greater stresses. As Richard Martin, author of several excellent articles on Atlas, pointed out, "Peery used his famous bulldog tenacity to convince management at General Dynamics and Lewis to test a full-scale Atlas-Centaur to the point of collapse."[36] Martin explained, "A bending moment about 80 percent greater than that at the outset of buckling was achieved before significant nonlinear deflections occurred, but there were some worrisome local buckles around protuberances like the liquid oxygen line outlet. Therefore, the analysis was adapted to allow only about a 60 percent increase in the applied load."

Collaboration began at this time between Martin and Gerus, leaders of dynamic analysis groups at General Dynamics and Lewis, respectively. These groups laid the foundation for an innovative pitch and yaw program that was later implemented when a new Teledyne digital computer was introduced after 1973. The Automatic Determination and Dissemination of Just Updated Steering Terms (ADDJUST) program allowed design loads from flight winds to be reduced by about 40 percent.

Silverstein insisted on a program review at Lewis once a month. Contractors were expected to spend two or three days at the laboratory. Vince Johnson, Centaur program manager from Headquarters, attended these meetings but was careful not to come between General Dynamics and its new Lewis managers. Johnson had the job of defending the program to Congress and running interference at Headquarters. In Bruce Lundin's view, Johnson "played the role of

[36] Communication to the authors by Richard Martin, 7 March 2002. See Richard E. Martin, "The Atlas and Centaur 'Steel Balloon' Tanks: A Legacy of Karel Bossart, "40th International Astronautics Congress paper, IAA-89-738, Cot 7-13m, 1989; "A Brief History of the Atlas Rocket Vehicle, Part III," *Quest—The History of Spaceflight Quarterly* 8 (2001): 48. See also David Peery, *Aircraft Structures* (New York: McGraw-Hill, 1949).

program manager absolutely perfectly."[37] His levelheaded approach to program management was one of the keys to the program's ultimate success.

Deane Davis, General Dynamics Centaur project manager, recalled that General Dynamics grudgingly admitted that Lewis people knew somewhat more than they did about liquid hydrogen. The old Centaur design had required the engines to be prechilled for a short time before Centaur engines were ignited in space. This wasted some of the precious liquid-hydrogen fuel. One of the project's important innovations was to "chill down" the vehicle on the ground, using liquid helium as a precoolant for the engine. This was a direct transfer of knowledge from Project Bee, which included precooling the engine. Liquid helium was a relatively rare chemical at that time, but it was much less difficult to handle than liquid hydrogen. Engineers at General Dynamics were skeptical that it could be procured in large enough quantities. Within minutes of leaving a contentious meeting, Silverstein was on the phone ordering a dewar of liquid helium from the government cryogenics laboratory in Colorado. Liquid helium precooling worked perfectly.[38] General Dynamics and the Lewis engineers also worked together on a new design for the insulation panels–the source of failure on the first launch.

During this time, General Dynamics engineers also solved the critical problem of the leaking of liquid hydrogen through minute pores in the welds of the liquid-hydrogen tank. Tests at General Dynamics revealed that metal became brittle when exposed to the very low temperatures of liquid hydrogen. It was found that by adding nickel in the weld area, the stainless steel could be strengthened. New techniques for lap-welding the seams, followed by spot welds, provided additional structural integrity. Factory technicians, already highly skilled builders of the Atlas tank, had to raise the standard even higher. The vehicle required more than 74,000 spot welds, 360 feet of resistance seam welds, and 400 feet of heliarc fusion butt welds. The welds were carefully monitored and x-rayed to make sure there were no imperfections.[39] In the most critical area of the design–the intermediate bulkhead–where cracks as small as 1/10,000 of an inch could destroy the vacuum created by freezing out the dry nitrogen gas in the space inside the double wall, the gores were fusion-welded.[40] Advanced welding techniques ensured a high degree of structural integrity without adding excessive weight.

Getting to the bottom of the problem of leaks in the intermediate bulkhead led to the discovery of another potential problem. Small cameras placed in the hydrogen tank revealed that as soon as the liquid hydrogen was loaded, the bulkhead mysteriously wrinkled. "Everyone took a look at those pictures and just about fainted, me included," Deane Davis recalled. "And

---

[37] Interview with Bruce Lundin by Virginia Dawson, 7 March 2000.

[38] John Sloop interview with Aerospace Division, Convair/General Dynamics, 29 April 1974, NASA Historical Reference Collection.

[39] Irwin Stambler, "Centaur," *Space/Aeronautics* (October 1963): 74–75.

[40] Ibid.

of course Abe Silverstein got all upset about it."[41] Testing determined that when the hydrogen was removed, the wrinkles disappeared. Through further testing and analysis, the team concluded that the stainless steel experienced "cryoshock" from the very cold liquid hydrogen. The way to avoid cryoshock was to load the liquid oxygen first, allowing the system to chill down gradually. Then the liquid hydrogen was slowly loaded. Wrinkles in the bulkhead no longer appeared. At the same time that they were solving these problems, engineers at General Dynamics and Lewis were building a formidable expertise in liquid hydrogen that they freely shared with Douglas and Rocketdyne, the contractors for the Saturn upper stages.[42]

New Regime at the Cape

Silverstein insisted on a major change in launch operations that had important implications for the future of robotic operations at the Cape. Under Marshall's management, responsibility for the rocket had ceased once it reached the launch pad. There, the Air Force director of launch operations took over. As part of the reorganization to facilitate Centaur development, the Air Force agreed to allow NASA to manage launch operations. Goddard Space Flight Center Launch Operations Branch at the Atlantic Missile Range had previously managed only the Delta launches. The branch, staffed by veterans of the Navy Vanguard Earth satellite program, now assumed responsibility for all robotic operations. (In 1965, all operations were consolidated under the John F. Kennedy Space Center, although former Goddard personnel continued to be in charge of robotic launches).

Headquarters placed Robert Gray in charge of the Goddard staff at the Cape. Gray had graduated from Allegheny College in Meadeville, Pennsylvania, with a degree in physics after World War II. After working at Bell Aircraft in rocket test instrumentation, he joined the Naval Research Laboratory in 1956. He served as launch director for Vanguard launches from the Cape, and when NASA took over, his group became part of Goddard. Gray named John Neilon, another Vanguard veteran, as his deputy. Neilon, a graduate of St. Anselm College in Manchester, New Hampshire, had also worked at the Naval Research Laboratory, where he evaluated the performance of precision location radars. Neilon would later take over as Director of Unmanned Launch Operations in 1970, when Gray became Deputy Director of Launch Operations at Kennedy Space Center. Gray also appointed John Gossett, another veteran of the Naval Research Laboratory, as chief of the Centaur Operations Division. In this capacity, Gossett acted as liaison with Lewis for Centaur. As time went on, a standard checkout procedure evolved. Gossett was responsible for checking in hardware shipments, scheduling

[41] John Sloop interview with members of the Aerospace Division Convair/General Dynamics, 29 April 1974, NASA Historical Reference Collection.

[42] Roger Bilstein, *Stages to Saturn* (Washington, DC: NASA SP-4206, 1980), 153, 188–189.

Technicians at Pad 36A gingerly lower Centaur into the interstage adapter on top of Atlas, August 1964. (KSC 64-16746)

tests, and "riding herd" on the contractors. He also worked with the range safety officers on documentation and launch scheduling.[43]

Because Silverstein strongly believed that the unusual technical challenges of liquid hydrogen required Lewis engineers to participate in the launch, he sent Bruce Lundin down to the Cape to make this clear to Kurt Debus, Director of the Launch Operations Center. Debus quickly agreed, presumably because he had little interest in robotic operations on the Air Force side of the Banana River.[44] The Goddard staff found working with hydrogen challenging. Gray said, "We spent years, really, and dozens of launches before we got all the strange little things that we didn't understand straightened out on that Centaur stage. Things like prestarts and in-flight ignitions and that sort of stuff. We had all these chilldowns to run through and all that stuff was very, very critical."[45] Preparing Centaur for launch was like fine-tuning a Jaguar sports car. "It had to be tweaked up all the time," Gray said, "and it was not tolerant to any kind of failures, or something not being just right. It had to be right, period, or else it wasn't going to work." Gray recalled that his launch team had to invent the launch rules to meet the unique demands of liquid hydrogen.

Mating Centaur with Atlas proved a difficult and delicate maneuver. The first time they gingerly lowered Centaur to sit atop the Atlas, the mechanical separation latches on the interstage adapter did not line up with the latch receptacles. These receptacles were large tabs welded to the vehicle with slots designed to receive the latching mechanisms. Red Lightbown, General Dynamics Senior Engineering Structures Manager for Atlas, was immediately summoned from San Diego. He discovered basic problems with the separation system and decided that the whole thing should be scrapped. A new system that used shaped charges was designed. The charge, placed inside a metal container attached to the interstage adapter, exploded with just enough energy to slice through the adapter, releasing Centaur from Atlas. To reduce the time it took to separate the two rockets from 45 seconds to 6, the retrorockets on Atlas were redesigned to provide more power.[46]

Karl Kachigan, a longtime Centaur chief engineer, ran the launch support team for General Dynamics. A graduate of Marquette University's School of Engineering, Kachigan had made his reputation during Atlas development. He produced a major paper in 1955 that described forced oscillations of a fluid in a cylindrical tank. Because Atlas had no rings inside the tank to dampen the oscillations of the liquid propellants, it needed baffles to prevent propellant sloshing. Kachigan's contributions to the problem of sloshing proved to be a major contribution to rocket theory.

During the five or six weeks before a launch, the Cape swarmed with engineers from General Dynamics and Lewis who worked closely with the Goddard robotic launch operations staff. At

[43] Interview with John Gossett by Virginia Dawson, 2 July 2002.

[44] Interview with Bruce Lundin by Virginia Dawson, 7 March 2000.

[45] Interview with Robert Gray by Virginia Dawson, 9 November 1999.

[46] Letter from Roger C. Lynch to Virginia Dawson, 4 June 2000.

that point, sleepy Cocoa Beach was not the vacation destination it is today. A few inexpensive motels adorned its expansive beaches. Small restaurants like Fat Boy's became hot spots where the launch teams could grab a meal of chili and hamburgers at any hour.

Kachigan's leadership helped shape the early General Dynamics Centaur culture. Several weeks before a launch, he sent a "tiger team" to the Cape. The members of the tiger team, key people from the design and technical groups,[47] were to "muck through" all the paper, look at all the hardware, and examine all the test procedures and analyze the data. They tried to find the slightest flaw that might compromise the launch. Kachigan headed investigations when a failure occurred. Tough, analytical, and fair, he was universally respected for his ability to ferret out problems and make decisions under pressure.

Because the vehicle could not tolerate the slightest failure of a single component, every system had to be checked out. Telemetry—the transmission of real-time data such as pressure, velocity, and surface angular position by radio link from the vehicle to the ground—was an important tool for assessing whether various systems were functioning properly. Through telemetry, the government could independently monitor prelaunch data to make sure that what the contractor told them was correct.

Once Atlas and Centaur were mated, each was inflated with nitrogen and held rigid with large suspension cables to prevent collapse. The Atlas-Centaur vehicle was then ready for extensive prelaunch testing. First, the tiger team ran the propulsion, guidance, and telemetry systems independently. Then they simulated a launch without propellants. All this testing culminated in the so-called tanking test, when liquid hydrogen and liquid oxygen were actually loaded. A simulated launch of the entire system, called a "wet dress rehearsal," verified that the valves were working correctly, pressures and temperatures were right, and all other systems worked in the low-temperature environment. NASA and General Dynamics carefully reviewed the data at a series of preflight readiness reviews. Whenever a problem in the data was found, the affected system had to be taken apart and studied, then put back together and tested.[48] As the launch day neared, a technician performed a coin-tap test, literally tapping the composite payload fairing and insulation panels of the Centaur to make sure no laminations of the skins had pulled away from the core and left air pockets. Then, shortly before declaring the vehicle ready, launch director Robert Gray took his ritual walk down the entire length of the vehicle and visually inspected it for any flaws that more sophisticated testing might have missed.

On launch day, about thirty engineers huddled in Building AE, where the atmosphere was thick with cigarette smoke and sweat. A launch erased the distinction between contractor and civil servant, with everyone focused on the job of checking and rechecking every system. Building AE

---

[47] Interview with Karl Kachigan by Virginia Dawson, 7 June 2000.

[48] Joseph Green and Fuller C. Jones, "The Bugs That Live at -423°," *Analog Science Fiction/Science Fact* 80 (January 1968): 30.

General Dynamics Launch Conductor Dan Sarokon (foreground) and Roger Lynch, Centaur Operations Manager, prepare for an Atlas-Centaur countdown, 1964. (Courtesy of Lockheed Martin)

housed the instrumentation support for all unpiloted launch vehicles. During the period between propellant loading and liftoff, all NASA engineers in the AE Building focused on telemetry data spewed out on a series of strip charts. If any of the data indicated an anomaly, an engineer would immediately inform the people in the blockhouse, who then might order the launch stopped.

Pressure was especially intense in the blockhouse, where Dan Sarokon, the General Dynamics launch conductor; Gray; Roger Lynch, Centaur launch operations manager; and representatives from Honeywell and Pratt & Whitney sat with about sixty other engineers. They checked the vehicle's vital signs on instruments that showed voltages, temperatures, and pressures. Sarokon pushed the button that began the final countdown. With Atlas-Centaur underway, the director of the NASA Lewis Launch Vehicles Division, along with other NASA engineers, sat glued to their monitors in Building AE. Telemetry data proved to be particularly valuable if a launch failed. Vanguard rocket veteran Skip Mackey, who served as head of the telemetry station for over forty years, provided the blockhouse with real-time commentary on the data by telephone. "All the years of preparation for a launch came down to the push of a button," one engineer remarked: "If it goes up you are a hero. If it doesn't, you're a goat."[49]

---

[49] Interview with Joe Nieberding by Virginia Dawson and Mark Bowles, 15 April 1999.

Left to right, Krafft Ehricke, the first Centaur Project Manager; Roger Lynch, Centaur Launch Operations Manager; Roger Lewis, President and Chairman of the Board of General Dynamics; Grant Hansen, the new Centaur Project Manager; and Ed Heineman, Corporate Director of Technology, wait tensely before the launch of AC-2 inside the blockhouse during a planned countdown hold, 27 November 1963. (Courtesy of Lockheed Martin)

Out of the Doghouse

It took eighteen months of intense preparation for the first test flight of Centaur under the new Lewis management. To underline the importance of the Centaur connection with Saturn, President John F. Kennedy visited the launch site as the launch date approached in November 1963. Several days later, Atlas-Centaur engineers and technicians were devastated to hear that the President had been assassinated in Dallas. NASA delayed the launch of Atlas-Centaur 2 (AC-2) one day to allow the Atlas-Centaur launch team to mourn Kennedy and to refocus on the arduous job of preparing for launch. Because no launch vehicle had ever required jettisonable insulation panels, there was no knowledge to apply in their design. Uncertainty over how they would function induced Silverstein to order them bolted to the Centaur tank in what he called a "brute-force fix."

First tension (bottom), then cheers (top) erupt in the control room after the first successful launch of AC-2, 27 November 1963.
(Courtesy of Lockheed Martin)

On 27 November 1963, the Atlas-Centaur lifted off from Cape Canaveral at 2:03 p.m. Atlas pushed Centaur to an altitude of 150 miles. The upper stage separated flawlessly from Atlas and fired its two 15,000-pound-thrust RL10 engines. This first flight proved the compatibility of Atlas with the upper stage Centaur and also proved that liquid-hydrogen engines could be fired in space. The same day, Pratt & Whitney Aircraft issued a classified report entitled "RL10 Engine for Advanced Space Missions."[50] The RL10, it announced, "is a mature engine with proven durability and reliability." The company's description emphasized the simplicity and controllability of the operating cycle.

On the heels of the Centaur launch came a test launch of the Saturn S-IV stage. It was powered by a six-engine cluster of RL10 engines with 90,000 pounds of thrust on 29 January 1964. This liquid-hydrogen upper stage had a more conventional reinforced structure with its insulation placed inside the rocket tank. The claim that Centaur was the first successful launch of a liquid-hydrogen upper stage was established just in time.

Although Centaur showed results, Headquarters agonized over whether the United States could win the race to the Moon. Lift capabilities of the United States still did not yet equal those of the Soviet Union. Deputy Administrator Seamans asked William Fleming, Director of Program Review, to compare the ability of the U.S. and U.S.S.R. to land unpiloted spacecraft on the Moon. Using intelligence reports, he concluded that the total weight the Soviet Union could land on the Moon in 1964 was 850 pounds, compared to a weight of 750 pounds for Surveyors to be launched in 1966. An improved Centaur, slated to be flightworthy in 1967, would be able to land a Surveyor payload weighing 1,000 pounds. He concluded, "It is felt that the Surveyor program, complemented by flights of the Lunar Orbiter beginning late in 1966 and followed by an Apollo manned landing in 1969 would equal or exceed any program of lunar missions that the Russians might accomplish over the corresponding period of time."[51] Fleming's memo led Homer Newell to strongly recommend the development of a Saturn 1 B-Centaur vehicle to cover the needs of larger lunar payloads and advanced planetary programs. Newell concluded, "Based on Apollo landings on the Moon in the late 1960's, the sequence of technological and scientific achievements represented by Ranger, Surveyor, Lunar Orbiter, and Apollo should offset many Soviet 'spectaculars.'"[52] However, just in case the Apollo program was delayed into the 1970s, Newell thought that NASA might consider sending a larger unpiloted lunar spacecraft to explore the lunar topography and return lunar samples to Earth.

Despite the reassurance he provided to Seamans, Newell was clearly dissatisfied with the progress of the Surveyor program. To William Pickering, Director of JPL, Newell emphasized the importance

---

[50] "RL10 Engine for Advanced Space Missions," Pratt & Whitney Aircraft, 27 November 1963, Box 59, Old RL10 Records (Goette files), DEB Vault, NASA Glenn.

[51] William Fleming, "Lunar Program Planning," 27 May 1964, John Sloop papers, NASA Historical Reference Collection.

[52] Homer E. Newell to Associate Administrator, 3 April 1964, John Sloop papers, NASA Historical Reference Collection.

of the Surveyor program to NASA. "Within NASA the Surveyor project is near the top of the overall priority list. Within the Office of Space Science and Applications, because of our commitment to provide the needed support to the Apollo mission, as well as because of its scientific value, Surveyor is at the top of our priority listing."[53] He did not think that JPL management was giving the program adequate support. He wanted closer monitoring of the contract with Hughes Aircraft for the spacecraft and better communication with Headquarters over Surveyor progress and needs.

During this time, Surveyor earned a well-deserved reputation with Congress as "one of the least orderly and most poorly executed of NASA projects."[54] Many of the problems stemmed from poor planning and lax management. Representative Joseph Karth, a vocal critic of the Centaur, admitted that Surveyor had not received enough scrutiny when it was conceived in early 1960: "We were in such a sweat to get going that we said, 'Let's not think about it, let's do it.' I was on the space committee then, and it's true, we were impetuous."[55] The cost-plus-fixed-fee contract allowed Hughes to run up huge project costs, charging them as NASA-directed changes without penalty. This was the first large space contract that NASA let to Hughes Aircraft. Hughes used it to learn the problems and pitfalls of building a spacecraft.

By 1964, the multiple problems associated with Surveyor development were so serious that NASA ordered a full review and concluded that the problems were both technical and managerial. Lack of supervision at Hughes, poor oversight by JPL, and neglect by NASA Headquarters all contributed to the inability of the Surveyor project to meet standards. For example, tests of the terminal descent guidance control system revealed problems. The vernier motors, designed by the Reaction Motor Division of Thiokol Chemical Company, were so undependable that JPL sought an additional contract with Space Technology Laboratories to provide a backup. This added more costs to the program.[56]

Under Lewis management, Centaur was equally plagued with a concatenation of problems that slowed development. AC-2's bolted-down insulation panels had carried instruments that revealed that the panels were not sturdy enough to withstand the aerodynamic forces Centaur encountered during flight through the atmosphere. The redesign increased the weight of Centaur, always a critical factor with respect to the amount of payload the vehicle could carry.

Ground tests revealed that the helium used to chill down the engines and purge the insulation needed to be released at a higher rate immediately prior to launch. In addition to cooling the engines prior to starting them in space, helium prevented the insulation from freezing to the skin of the liquid-hydrogen tank. It was decided that an additional helium tank needed to be

---

[53] Homer E. Newell to William Pickering, 13 July 1964, Surveyor files, NASA Historical Reference Collection.

[54] H. R. Committee on Science and Astronautics Investigation, 1965. See also Clayton Koppes, *The JPL and the American Space Program*, 173.

[55] Ibid, 173.

[56] Ibid, 173–183.

carried on board to continue the helium purge during liftoff.[57] The next Centaur (AC-3), launched in June 1964, jettisoned the insulation panels and nose fairing successfully, but, almost immediately, a freak failure of the hydraulic system caused the vehicle to spin out of control. Then AC-4, launched in December 1964, tumbled wildly during the coast phase.

Through a massive effort that involved both analysis and testing, engineers traced the cause of the AC-4 failure to uncontrolled propellant behavior after venting. Fred Merino of General Dynamics and Lewis staffers Andy Stofan, Steve Szabo, and Ray Lacovic conducted some of the first investigations of zero-gravity propellant management. They initiated a test program using the 100-foot drop tower at Lewis for experiments. Later, an elaborate, state-of-the-art, 500-foot zero-gravity facility was constructed for basic research on liquid hydrogen. Merino marveled at the openness of Lewis engineers. He often went to Lewis for a month to observe tests. "There were times when I would come up with a problem," he said, "and we would have to relate it to the people in Cleveland and give our rationale for why we thought there was a need for a design change. And there were many times that they would run tests for us. It was their test, but we provided input as required to get the results that were needed to effect the design change to meet a mission requirement."[58] With this type of creative collaboration, program personnel made rapid progress on resolving problems.

The test program focused on the problem of the sloshing of propellants in flight—information needed for designing the autopilot controls. Through tests and analysis, scientists discovered that in zero gravity, liquid hydrogen adheres to the sides of the tank with the ullage area (space where the gas vapors collect) in its center. Because of the position of the ullage area, a large amount of liquid-hydrogen fuel had been released, along with hydrogen gas, when AC-4's vent valve opened. As the liquid flowed overboard, it expanded with a tremendous amount of evaporation and formed a ball of gas. This ball exerted a force on the vehicle that caused it to tumble. Merino and colleague Clay Perkins developed a patented vapor sensor to signal to the system's computer whether the valve was surrounded by liquid or vapor, and they redesigned the vent system. This knowledge was shared with engineers designing a propellant-management system for the liquid-hydrogen stages for Saturn. Confidence in liquid hydrogen as a propulsion fuel was building.

The new knowledge of the ullage area also led to greater confidence in managing the propellants during the coast phase. A Reaction Control System (RCS) was designed to provide acceleration just before venting. The system consisted of a hydrogen peroxide supply bottle, lines, valves, and small vernier motors called thrusters. Before the tank pressure reached a certain point, small thrusters in the tank were fired to reposition the propellants in the aft end of the vehicle. Then a valve could be opened to relieve pressure without danger of loss of liquid

[57] See Green and Jones, 27.

[58] Interview with Fred Merino, 5 June 2000. See also Centaur Project Office, "Coast-Phase Propellant and Vehicle Behavior," paper presented to Low Gravity Cryogenic Fluid Mechanics Ad Hoc Committee Meeting, Convair/General Dynamics, 16–18 March 1965.

propellant. The same thrusters were fired in preparation for starting the engine, so that liquid rather than gaseous hydrogen and oxygen entered the pumps.

Continuing Doubts

Engineers were confident that the launch of AC-5 in March 1965 would succeed because direct ascent was not as complicated. The checkout went forward without incident. Countdown was equally flawless. The Atlas engines fired, the hold-down pins retracted, and Atlas-Centaur lifted off. At 5 feet off the ground, the Atlas engines suddenly shut down and the vehicle dropped back to Earth, causing an explosion that extensively damaged the pad. An eyewitness provided this understated report: "The burning gas cloud had an expansion velocity of over 3,000 feet/second, and furnished one of the best examples seen at the cape of why fragile human beings retreat to massively built blockhouses during a launch. No one was injured, other than feelings, but AC-5 was a mass of blackened rubble, and Pad A had suffered major damage and would require extensive rebuilding."[59] Failure analysis revealed that after liftoff, an improperly installed instrumentation pressure transducer on Atlas had caused a valve to close, cutting off fuel to the engines. Although it had nothing to do with Centaur, this explosion secured Centaur's reputation as the "hard luck bird" once and for all.[60]

The NASA Budget Authorization Hearings held the same month gave Representative Joseph Karth the opportunity to condemn General Dynamics for an "inordinately bad job" on Centaur.[61] Congress was unhappy about the enormous costs of the program—now between $600 and $700 million. The price of the rocket seemed to reflect NASA's dependency on General Dynamics as a sole-source supplier. Implicit in Karth's aggressive questioning was the issue of the unique design of Centaur—the thin-skinned balloon structure and the intermediate bulkhead, in particular. Previously, when Centaur had run into troubles, NASA had attributed them to the difficulties of developing a liquid-hydrogen rocket. Now Marshall had tested the S-IV stage for the Saturn C-1 with the reinforced design favored by von Braun. Marshall engineers had also developed a new Rocketdyne J-2 engine designated for the second and third stages of Saturn V. This engine burned liquid hydrogen and produced 150,000 pounds of thrust, compared to the RL10 engine's 15,000 pounds.[62] Were the foibles of pressure-stabilized design, rather than the diffi-

---

[59] See Green and Jones, 30.

[60] See Green and Jones, 31. See also William Hines, "Atlas Blast Laid to 'Impossible' Mishap," *Evening Star* (5 March 1965).

[61] Subcommittee on Space Science and Applications, *1966 NASA Authorization*, Hearings, H.R. 3730, March 1965, Part 3, 1143–1144. NASA had also planned to use Centaur as an upper stage for the Saturn 1-B to be flown on missions after 1968; see p. 1138. Originally, in addition to the eight development flights for Centaur, seventeen flights of Surveyor were planned, plus six for Saturn 1-B in support of Voyager. The Saturn 1-B/Centaur program was terminated in 1968.

[62] On the J-2, see Roger Bilstein, *Stages to Saturn* (Washington, DC: NASA SP-4206, 1980), 140–153; Paul N. Fuller and Henry M. Minami, Jr., "Rocketdyne Reborn," *Space* 3 (1987): 55–58; W. R. Studhalter, "The J-2 Liquid Hydrogen Rocket Engine," National Aero-Nautical Meeting, Society of Automotive Engineers, April 1963.

culties associated with its liquid-hydrogen propellant, holding back Centaur development? Karth thought General Dynamics was the real problem and cited past problems with the intermediate bulkhead as evidence of a poor job. "This has been a very favored program. It has, if you please, had a DX priority for 3 years, the highest possible. Tremendous effort has gone into it."[63] He thought that NASA should consider an alternative source, such as Titan III with a high-energy upper stage. At this time, JPL was agitating for launching Surveyor on the Titan III with an upper stage called the Transtage. Congressman Weston E. Vivian of Michigan supported the idea of creating some competition among aerospace companies:

> It would seem to me you have a program heavily dependent on Centaur, specifically in the unmanned spacecraft area. You are planning procurement of several hundred million dollars more equipment of this variety. You have a single contractor on whom your hopes rest totally, and whose dollar estimates will control your dollar estimates. I would frankly feel a competitive program should be initiated, and even though the apparent cost initially might be higher, I would certainly consider that you should open bids for procurement of a given number of Centaur vehicles, say one-third of the total you are planning to buy, from an alternate source.[64]

In response, Homer Newell reaffirmed NASA's commitment to General Dynamics and the company's unique expertise in manufacturing the thin-skinned structure: "The capability of building thin-wall pressure stabilized tankage to our extremely high standards has been highly developed at Convair and is not currently available elsewhere. Because of these considerations, DOD and NASA have concluded that it is not practicable to consider alternate sources for either Atlas or Centaur."[65]

This congressional pressure for more competition was not lost on Martin Marietta-Denver. The president of the company, I. Nelvin Palley, had already offered its Titan IIIC to NASA as an alternative to Atlas-Centaur. The company had also raised this possibility with DOD and interested members of Congress.[66] After the first flight of Titan IIIC proved successful in July, the Air Force offered to carry Surveyor to the Moon on Titan's next flight, scheduled for September.[67] No mention was made of the fact that Surveyor was far from being ready.

---

[63] Subcommittee on Space Science and Applications, *1966 NASA Authorization,* Hearings, H.R. 3730, March 1965, Part 3, 1143–1144.

[64] *1966 NASA Authorization,* Hearings, 1140.

[65] *1966 NASA Authorization,* Hearings, 1147.

[66] Earl Hilburn to Robert Seamans, Jr., 22 March 1965, Surveyor files, NASA Historical Reference Collection.

[67] "Titan 3C Offered to Soft Land Surveyor," *Cocoa Tribune* (21 June 1965). See also William J. Normyle, "Congress May Force NASA Titan 3 Study," *Aviation Week & Space Technology* (12 July 1965): 30.

The spectacular explosion of AC-5 in March 1965 left Pad 34A a mass of blackened rubble. (NASA Glenn Research Center unprocessed photo)

Because of the damage to the launch pad from the explosion of AC-5, launch complex 36B was rushed into service, making the launch of AC-6 possible on 11 August 1965. This was the last R&D flight of the direct-ascent program. The Atlas-Centaur operated flawlessly, successfully injecting a dummy of the Surveyor spacecraft into a prescribed transfer trajectory and simulated impact with the target, the so-called "paper Moon." AC-6 flight-tested an improved propellant control system. Cleveland's *Plain Dealer* headline announced, "Paper Moon Shot Gets Rocket Out of Doghouse."[68] The article pointed out that the Russians had failed twice that year in soft-landing attempts, then added that three of the previous five tests of Centaur had also failed. The feasibility of Centaur was still a matter of opinion. In response to inquiries from the ever-vigilant Representative Karth, Seamans responded defensively:

> The slippages in schedule and the increases in program cost are serious problems; but it must be realized that this mission is one of the most difficult technical assignments we have undertaken in the exploration of space. We are continuing to do everything we can to maximize the chances of successful engineering flights in 1966 and of scientific missions in 1967. To accomplish these objectives, it may become necessary to reprogram additional funds in FY 1966 for the Surveyor project.[69]

The relentless criticism by the press and Congress continued. In March of 1966, the Committee on Aeronautical and Space Sciences again complained that generous funding for both Centaur and Surveyor had produced less than spectacular results. Legislators questioned whether the Apollo program's need for Surveyor had passed. In his confidential testimony before the committee, Newell responded that NASA still regarded Surveyor as important for verifying the design concepts used in the Apollo Lunar Excursion Module (LEM). The first Surveyor would also provide information on landing loads at surface impact. Its TV camera would locate protuberances and depressions in the lunar terrain, areas to avoid when selecting an appropriate landing site.[70]

What Surveyor could no longer guarantee was the quality of the science it would return. Narrow Apollo objectives had compromised plans by scientists to use Surveyor to acquire extensive knowledge of lunar geography and geology. Ronald Scott, a scientist at Caltech, complained that the scientists recognized that Surveyor "is now essentially a back-up mission for Apollo."[71] Milwitzky agreed that the "massive difficulties in getting the basic Surveyor system to work" had drained both manpower and attention away from the scientific missions

---

[68] *Plain Dealer*, 12 August 1965.

[69] Robert Seamans, Jr., to James Webb, 10 November 1965, John Sloop papers, NASA Historical Reference Collection.

[70] Confidential testimony, Homer Newell, 2 March 1966, John Sloop papers, NASA Historical Reference Collection.

[71] Ronald Scott to Benjamin Milwitzky, 22 December 1965, Surveyor files, NASA Historical Reference Collection.

and changed its basic goals. If Apollo remained on schedule, the role of Surveyor would be limited to the determination of the Moon's surface. Should it turn out to be a surface that would support a lunar landing, this would be an important contribution in view of the Apollo program's national importance. If they were to discover that the surface was relatively soft, more flights of Surveyor might be necessary to provide additional data. Milwitzky speculated that the Apollo mission might be delayed sufficiently to allow the scientific flights to get to the Moon first. He reassured Scott that the science missions were an essential part of NASA's lunar exploration program: "there is no intent on our part to cancel these missions or let them slowly wither away on the vine."[72]

Although Centaur was ready for a direct shot to the Moon, Surveyor was still mired in development problems. Silverstein decided to launch another R&D flight to test Centaur's restart capability. Unfortunately, AC-8, launched in April 1966, did nothing to dispel the concerns of the scientific community and Congress. A leak in the hydrogen peroxide system used to settle the propellants made it impossible to start one of the Centaur main engines after the coast phase. This unbalanced thrust sent the vehicle into a tumble. *Space Daily* dryly reported, "The Centaur still has not demonstrated its parking orbit capabilities, the absolute requirement to justify the development of the Centaur transportation system."[73] Yet to the engineers involved in the redesign of the baffles and vents, the system was finally satisfactory. The leak in the hydrogen peroxide system had nothing to do with the restart capability. They were convinced that the restart problem was solved.

To increase its leverage with Pratt & Whitney, Lewis Research Center was finally able to take control of management of the RL10 engine from Marshall in May. A new office, managed by William Goette with a staff of four, was set up. The office would be responsible for testing the more advanced RL10A-3-3, qualified in November 1966.

A Soft Landing on the Moon

At the end of May, Surveyor appeared ready for a first flight, although NASA acknowledged that a large number of last-minute changes to Surveyor equipment increased the risk of failure. Although the mission might not achieve its primary objectives, it would yield enough useful data "to permit greater confidence in overcoming some of these uncertainties."[74] After a nearly flawless launch on a bright Memorial Day morning, AC-10 hurtled skyward carrying the first Surveyor. The Centaur RL10 engines ignited 4 minutes 12 seconds after liftoff. They operated

---

[72] Benjamin Milwitzky to Ronald Scott, 17 January 1966, Surveyor files, NASA Historical Reference Collection.

[73] *Space Daily* (11 April 1966), Centaur files, NASA Historical Reference Collection.

[74] Earl D. Hilburn to Dr. Seamans, "Special meeting to ascertain launch readiness of Surveyor," 25 May 1966, Surveyor files, NASA Historical Reference Collection. For a technical overview of the Surveyor Program, see J. Jason Wentworth, "A Survey of Surveyor," *Quest* (Winter 1993): 4–16.

The Fourteen Launches of the Atlas-Centaur for the Surveyor Program (1963–1968)				
Launch Date	**Mission**	**Vehicle**	**Payload**	**Objective**
27 November 1963	R&D (one-burn)	AC-2	No payload	Demonstrate separation.
30 June 1964	R&D (one-burn)	AC-3	No payload	Demonstrate jettison of insulation panels and nose fairing. Failed.
11 December 1964	R&D (two-burn)	AC-4	2,070 lb	Demonstrate coast-phase propellant control.
2 March 1965	R&D (one-burn)	AC-5	1,411 lb	Demonstrate operational readiness. Failed.
11 August 1965	R&D (one-burn)	AC-6	2,100 lb	Demonstrate operational readiness in direct ascent.
7 April 1966	R&D (one-burn)	AC-8	1,730 lb	Demonstrate coast-phase propellant control. Failed.
30 May 1966	Surveyor 1 (one-burn)	AC-10	2,193 lb	First operational Atlas-Centaur flight. First controlled unpiloted landing on the Moon. First pictures from the Moon.
20 September 1966	Surveyor 2	AC-7	2,204 lb	Second soft landing on the Moon. Postlanding TV survey.
26 October 1966	R&D (two-burn simulated lunar transfer)	AC-9	1,740 lb	First engine restart of LH_2/LO_2 engines. Centaur operational for two-burn missions.
17 April 1967	Surveyor 3 (two-burn)	AC-12	2,281 lb	Perform soft landing within Apollo landing zone. Manipulate lunar surface with soil sampler.
14 July 1967	Surveyor 4 (two-burn)	AC-11	2,295 lb	Perform soft landing at Sinus Medii. Conduct vernier engine experiment.
8 September 1967	Surveyor 5 (two-burn)	AC-13	2,217 lb	Perform soft landing on Mare Tranquillitatis. Determine chemical elements in soil.
7 November 1967	Surveyor 6 (two-burn)	AC-14	2,220 lb	Perform soft landing at Sinus Medii.
7 January 1968	Surveyor 7 (two-burn)	AC-15	2,289 lb	Perform soft landing at Tycho. Last Surveyor mission.

General Dynamics launch team members at Kennedy Space Center (left to right): Tom Henry, Roger Lynch, and Deane Davis hear the good news of the successful touchdown of Surveyor 1 on the Moon, June 1966. (Courtesy of Lockheed Martin)

for 7 minutes 18 seconds before engine cutoff and release of the spacecraft. Then engineers and scientists 3,000 miles away at JPL in Pasadena, California, took over. They monitored data in the JPL Space Flight Operations Facility (SFOF), where there were banks of computers for analyzing trajectories and telemetry data and sending commands to the spacecraft. Their connection to the Deep Space Information Facility kept JPL engineers in radio contact with tracking stations around the world. The next day, the spacecraft made what was referred to as a "mid-course correction" to place it within 9 miles of its target. Cary Nettles recalled that what the Surveyor actually performed was not a mid-course correction, but reverse and forward firings to dump extra fuel before the craft landed–the result of a guidance miscalculation by JPL. He said:

> The tracking stations were indicating that the Centaur guidance had performed perfectly and the spacecraft was right on target. This situation came as a complete surprise to JPL and they needed to do something to get rid of extra fuel they had loaded on for their own estimate of what the guidance was going to do.[75]

---

[75] Letter from J. Cary Nettles to Virginia Dawson, 15 June 2002.

No one had expected both Centaur and Surveyor to operate flawlessly on the first try. The team at JPL scrambled to coordinate the landing as the spacecraft neared the Moon after traveling 230,000 miles in 63 hours. Its braking rockets slowed it from nearly 6,000 miles per hour to a bare 3.5 miles per hour. Then, 14 feet from the lunar surface, the rockets cut off and the vehicle dropped to the Moon's surface. All three footpads hit the surface at almost the same time and penetrated to a depth of a few inches. The spacecraft rebounded once before coming to rest about 35 miles north of the crater Flamsteed in Oceanus Procellarum on 2 June 1966. After 40 minutes of suspense, jubilant cheers sounded at JPL, Hughes, General Dynamics, Lewis Research Center, and Kennedy Space Center when the first pictures were flashed back. Silverstein's decision to go with direct ascent was vindicated.

Surveyor operated for two lunar days (about twenty-eight Earth days) before its batteries went dead. During this time, Surveyor transmitted about 15,000 striking pictures of the solar system, including the Sun's corona out to at least three or four solar radii and other celestial bodies including Sirius, Canopus, Jupiter, and Gemini. As the lunar night closed in, Surveyor recorded the rate of temperature decay. In their final positioning, Surveyor's solar panels faced west so that the craft would cast a long shadow at lunar dawn for the Apollo astronauts to see during their forthcoming mission.

The performance of Centaur made the Surveyor mission possible. Placed in its trajectory by Centaur with near-pinpoint accuracy, Surveyor hit the Moon surprisingly close to its target. A wire from Deputy Administrator Seamans congratulated the Lewis team. Referring to the transfer of Centaur to Cleveland three and a half years earlier, Seamans wrote, "The achievement of the Centaur organization, both government and contractor, is particularly striking when the technical difficulty and complexity of the Centaur development is properly understood. At times, some of these difficulties must have appeared to be nearly insurmountable. However, the outstanding technical competence of the people and the dedication which they have given to the Centaur project seems to have reduced all these difficulties to manageable size."[76]

The United States had achieved its first soft landing of a spacecraft on the Moon, but the Soviet Union's Luna IX had already beaten Surveyor to the Moon by over three months. Far simpler mechanically than Surveyor, Luna IX had a crushable structure that was damaged on impact. Thus, NASA could claim that Surveyor's landing was the world's first *controlled* soft landing. From the few pictures that Luna IX sent back, scientists were relatively sure that a spacecraft would not sink into lunar dust, but Surveyor's thousands of pictures were sharper and clearer. Gone were doubts about whether the surface could support the weight of the Lunar Excursion Module. Because Surveyor was intended to demonstrate the soundness of the basic

---

[76] *Lewis News* (10 June 1966); *The Orbit* (24 June 1966) (Pratt & Whitney Aircraft, Florida Research and Development Center).

landing technique the LEM would use, astronauts Deke Slayton, Edwin (Buzz) Aldrin, Thomas Stafford, and Eugene Cernan witnessed the first Surveyor's touchdown from JPL's control room.

The next launch—AC-7 carrying Surveyor 2, an engineering test model—was launched from the Cape in September 1966. This time, Surveyor's vernier engines failed, causing it to tumble out of control. It crashed on the lunar surface at a velocity of 8,737 feet per second. Despite this setback, Centaur's increasing reliability enabled NASA to justify an order of five additional Centaurs from General Dynamics on 23 September 1966. This time, the new contract for $15,565,331 was fixed-price, a controversial but astute way to control Centaur's costs. For their superhuman effort in turning the troubled rocket into a reliable upper stage, Cary Nettles and Ed Jonash (who took over as head of the Launch Vehicles Division when David Gabriel left) received NASA's Distinguished Service Medal in early October as they readied AC-9 for launch.

The next Centaur, launched on 26 October 1966 from Complex 36B, definitively proved Centaur's capability to restart its liquid-hydrogen engines in space. The vehicle carried a 1,750-pound dummy model of the Surveyor spacecraft. Centaur was able to compensate automatically for the shutdown of Atlas' sustainer engine 6 seconds early. Boosted into a 90-nautical-mile parking orbit by its first burn, AC-9 coasted in a circular orbit under low-gravity conditions for 24 minutes. The main engines of Centaur ignited, propelling the vehicle into the proper simulated lunar intercept trajectory. Although the Agena upper stage had proved its restart capability five years earlier, the ability to restart liquid-hydrogen engines in space was hailed as an outstanding achievement. With a coast period, Centaur launches were no longer restricted to the summer months but could take place throughout the year. The restart capability also increased the amount of available launch time each day.

The flight also verified the reliability of the inertial guidance system to control an indirect-ascent mission to the Moon. Homer Newell wrote to Abe Silverstein to congratulate the Center on this achievement: "Your Center may take justifiable pride in the aggressive manner in which this most difficult task was undertaken and the professional manner in which it has been completed."[77] Edgar Cortright's congratulatory letter stated:

Recalling the situation at the time you undertook to manage Centaur, I cannot think of a single other group in the country which would have accepted that responsibility under such dire circumstances. I can still recall your personal response to me when I asked you to consider it—which was, in effect, 'Someone has to do it.' Having watched Centaur development fairly closely, I am acutely aware of the personal sacrifice made by many Lewis employees, not to mention their families, in completing this development. Although it has been tough, I can't help but feel that most of them will look back

---

[77] Homer E. Newell to Abe Silverstein, 10 November 1966, John Sloop papers, NASA Historical Reference Collection.

on these years in the way many of us look back on our military service—namely, they hope they never have to do it again, but they wouldn't have missed it.[78]

Newspapers jumped on the bandwagon. "A powerful Atlas-Centaur rocket shattered a long-time scientific jinx today and handed the United States new muscle to lift heavy payloads to the Moon and planets," reported the *Chicago Tribune* on October 27. NASA hailed Centaur as the "world's first operational rocket using liquid hydrogen and liquid oxygen for propellants."[79]

Behind the scenes, however, analysis of the data from this flight revealed some disquieting facts. Besides the premature shutdown of the Atlas sustainer engine, one of the Pratt & Whitney engines had experienced an unexplained drop in performance. Although it had not affected the outcome of the mission, it was a bad sign. Another problem was that the hot exhaust from Centaur's nozzles was impinging on the turbine pumps, causing them to warm more than they should. As a precaution, protective shields were added to prevent damage from exhaust gases.

Surveyor III (AC-12), launched in April 1967, marked the first operational use of Centaur's restart capability—once for initial boost after Atlas separation and again to launch the spacecraft into a lunar trajectory after a short parking orbit that lasted approximately 25 minutes. After AC-12 released Surveyor, the spacecraft received a command to lock onto the Sun and Canopus, the largest star in the southern hemisphere. Surveyor coasted until it reached the point for its mid-course correction. Then one of the tracking stations sent the command to orient the spacecraft into its landing position. It landed only 3 miles from its planned target—a point on a moving object 237,000 miles away from Earth. The spacecraft took several unplanned, very large hops before coming to rest at an angle inside a lunar crater strewn with rocks on the Ocean of Storms. Surveyor III sent back thousands of pictures, including a picture of its own footprints. In addition to a camera, Surveyor carried a "soil scratcher" to scoop up lunar soil and break lunar rocks.[80] By examining photographs of the digging operations of its tiny scoop, scientists were able to calculate the bearing strength of the Moon's surface.

Surveyor V (AC-13), launched in September 1967 on an improved Atlas, landed on the Sea of Tranquility, 18 miles from its target. Experiments using an alpha scattering spectrometer, a device that bombarded the surface of the Moon with subatomic particles, proved that the lunar surface consisted of volcanic rock basalt with high titanium content. This was similar to soils found on Earth. Surveyor V sent back 18,006 pictures of terrain under consideration for the landing site for Apollo. Gone were any doubts that a controlled landing on the Moon could be accomplished. Surveyor VI (AC-14) explored another potential landing site and confirmed the

---

[78] Edgar Cortright to Abe Silverstein, 28 October 1966, and *Chicago Tribune* article, both reproduced in Centaur's *Tenth Anniversary* (Cleveland, OH: limited edition, 1972).

[79] *1968 NASA Authorization,* Hearings, March 1967.

[80] *Lunar Landmarks,* Surveyor Project Newsletter (24 April 1967), NASA Historical Reference Collection.

Apollo astronaut Charles Conrad, Jr., examines Surveyor III (landed 19 April 1967) with the Lunar Module 600 feet away on the Ocean of Storms, 1969. (NASA AS12_48_7136)

composition of the lunar soil. Then the vernier engines were ignited and the spacecraft "flew" to another location about 8 feet away.

By the seventh and final launch of Surveyor (AC-15) in January 1968, NASA had already selected all the landing sites for Apollo. Since Apollo support was superfluous, NASA decided to make the last Surveyor a completely scientific mission. The Surveyor team selected a site near Crater Tycho for the spacecraft's long arm to scratch and dig. Geologists hailed the results. Despite its many troubles, the Surveyor program, managed by JPL, proved to be a triumph for NASA and a highlight in the careers of the people involved in this ambitious undertaking. In reflecting on the significance of Surveyor, Oran Nicks wrote:

> It was as recently as 1966 that Surveyor 1—a robot that, on its first flight, flawlessly made a radar-controlled landing on another body in the solar system, and turned its television eye to stare about with the insatiable curiosity of its creators—showed that men setting foot on the Moon would not sink in a quick sand of Moondust. Since then we have learned that our expendable machines can be asked to perform tasks of remark-

able delicacy, and that our astronauts can be sent off with higher confidence of safe return than the hardy sailors who had first crossed the unknown oceans.[81]

Beyond demonstrating that the surface of the Moon was hard enough for the astronauts to land safely, Surveyor proved the usefulness of robotic spacecraft to perform complex tasks. Flight controllers learned how to repair spacecraft by remote control. Surveyor also served as a prototype for more advanced spacecraft, such as the Viking Mars lander.[82]

Even more important, but far less heralded, the mission's upper stage proved the feasibility of liquid-hydrogen fuel for transporting delicate spacecraft to a designated point in space. The RL10 engine was the first liquid-hydrogen rocket engine to be successfully restarted in space (October 1966). It proved that with a coast and restart, a spacecraft could be positioned for an optimal trajectory to another body in the solar system. With knowledge gained from Centaur, the designers of Saturn's J-2 engine for the upper stages of the Saturn V could move forward with greater confidence. Among the most significant contributions to basic knowledge of liquid hydrogen were the solutions to propellant sloshing and the venting of hydrogen gas in zero gravity.

Silverstein's advocacy of liquid hydrogen and his leadership proved to be crucial not only for the success of the Surveyor program, but also for the development of the Saturn rocket. Referring to the first launch of a Saturn V on the Apollo 4 mission on 9 November 1967, von Braun credited Silverstein with "pioneering work in liquid hydrogen technology [that] paved the way to today's success."[83] This was not an overstatement because Silverstein's unflagging advocacy had won von Braun's support for liquid hydrogen upper stages for the Saturn V. When Apollo 4 demonstrated the restart capability of the J-2, this decision was vindicated.[84]

At many points in *Stages to Saturn*, historian Roger Bilstein refers to the knowledge of liquid hydrogen that the designers of the Rocketdyne J-2 engine and the Douglas S-IV and S-IV B stages received through contact with the engineers working on Centaur. Douglas designers were able to benefit from many of the concepts pioneered on Centaur because NASA required its contractors to share their know-how with competitors.[85] For example, Rocketdyne engineers tried to use a familiar flat-faced copper injector design for the J-2, but the high operating temperatures of liquid-hydrogen/liquid-oxygen propellants caused it to burn out. They resisted adopting the porous face injector design of Pratt & Whitney's RL10 engine until a demonstration of its unique

---

[81] Oran Nicks, ed., *This Island Earth* (NASA SP-250, 1970), 166–167.

[82] Jason B. Wentworth, "Survey of Surveyor," *Quest* 2 (Winter 1993): 4–16.

[83] See signed photo in Virginia Dawson, *Engines and Innovation*, 194.

[84] Roger Bilstein, *Stages to Saturn*, 359.

[85] See Roger Bilstein, *Stages to Saturn*, 188–189; and Oswald H. Lange, "Development of the Saturn Space Carrier Vehicle," in Ernst Stuhlinger et al., *Astronautical Engineering and Science from Peenemünde to Planetary Space*, 21.

Surveyor's Cold War context is reflected in a cartoon from the Philadelphia Evening Bulletin *(22 April 1967): 135. (Courtesy of the Urban Archives, Temple University, Philadelphia)*

Rigi-Mesh injector face at Lewis Research Center convinced them. The injector proved to be the solution to their problem.[86] In considering the type and amount of insulation required to maintain tank temperatures at -432°F, S-IVB designers at Douglas consulted with people in the Convair Division of General Dynamics. However, after thoroughly studying the Centaur external insulation, they adopted the less technically complex but heavier solution of placing the insulation inside the tank.[87] Douglas and Rocketdyne engineers were forewarned of the propensity of liquid hydrogen to leak through minute pores in metals. They also learned precious lessons on venting and settling propellants prior to a restart in space. Bilstein wrote:

> Although the liquid hydrogen engines were developed and built by two different contractors, the government managed both programs so that information from one program was available to subsequent programs. Lewis Research Center, NASA's facility in Cleveland, represented an interesting intermediary influence, providing a pool of knowledge about liquid hydrogen technology used by Pratt & Whitney and Rocketdyne alike. Just as early work at Lewis was a benefit to Pratt & Whitney's RL-10, Rocketdyne's later J-2 benefited from both Pratt & Whitney and Lewis.[88]

This expertise in liquid-hydrogen technology might have been lost had Webb decided to yield to pressure for the cancellation of Centaur. In contrast to the American success with liquid hydrogen, the Russian rocket designer Sergei Korolev had constantly urged the Soviet government to support the development of liquid-hydrogen/liquid-oxygen upper stages but never received adequate funding. Author Asif Siddiqi makes the point several times in *Challenge to Apollo: The Soviet Union and the Space Race* that the Soviet Union had neither the technical expertise nor the manufacturing resources in the early 1960s to develop a liquid-hydrogen rocket. It also lacked adequate ground-test facilities. Siddiqi considered the failure to develop liquid-hydrogen upper stages "a strategic mistake that cost the Soviet space program much in terms of capability and efficiency, but Korolev alone did not have the force to single-handedly create a new industry in the Soviet Union."[89] Siddiqi points out that the first ground test of a Soviet liquid-hydrogen/liquid-oxygen engine was not carried out until April 1967. The Soviet Union was at least six years behind the United States in developing what he considered a "critical area of rocket engine technology."[90]

---

[86] See Roger Bilstein, *Stages to Saturn,*145.

[87] Ibid., 173.

[88] Ibid., 153.

[89] Asif Siddiqi, *Challenge to Apollo*, 318. A similar point is made by Tom D. Crouch in *Aiming for the Stars: The Dreamers and Doers of the Space Age* (Washington, DC: Smithsonian Institution Press, 1999), 179–180.

[90] Siddiqi, *Challenge to Apollo*, 548. Siddiqi notes on p. 840 that the actual launch of the Soviet liquid-hydrogen/liquid-oxygen rocket did not occur until May of 1987.

Left to right: Abe Silverstein; Fred Crawford, President of TRW; and former NASA Administrators T. Keith Glennan and James Webb at the dedication of the Glennan Space Engineering Building at Case Institute of Technology, January 1969. (Courtesy of Rebman Photo Service, Inc.)

On the tenth anniversary of the transfer of Centaur from Huntsville to Cleveland, Grant Hansen reminded Bruce Lundin, then Director of Lewis Research Center, of how close Centaur had come to being canceled. He regarded the transfer to Lewis as the key decision in saving the program. Reveling in the role they had played in pioneering the development of the world's first liquid-hydrogen rocket, he wrote, "Were it not for its being overshadowed by the Moon landing program, I believe that the Centaur story would have been the prime technical achievement of the last decade."[91] Homer Newell linked Centaur with the end of the anxiety over America's launch vehicle capability. He wrote, "By 1966–when Centaur became fully operational–the United States could at last launch spacecraft for just about any space mission . . . . Although the debate over whether the United States could or could not match Russian launch capability still arose occasionally, the subject no longer had the importance that it once did."[92]

---

[91] Letter from Grant Hansen to Bruce Lundin, 7 October 1972, reproduced in *Centaur's Tenth Anniversary* (Cleveland, OH: limited edition, 1972).

[92] Newell, *Beyond the Atmosphere*, 140.

By the late 1960s, the Civil Rights movement and President Lyndon Johnson's War on Poverty had turned national attention away from the imperatives of a space race with the Soviet Union. The war in Vietnam sapped the energies of the Johnson administration, leading to the President's decision not to run for another term. At the same time, James Webb, NASA's capable Administrator, lost President Johnson's ear. He had kept alive Kennedy's dream of landing human beings on the Moon, but he resigned before the historic flight of Apollo 11 in August 1969.

Krafft Ehricke still firmly believed in industrialization of the Moon and piloted flight to the planets. However, in the face of environmentalists and zero-growth advocates of the early 1970s, his enthusiasm for space tourism, orbiting hotels, and zero-gravity sports may have seemed more eccentric than prophetic. He continued to promote space travel as a means of transcending the petty divisions among human beings on Earth, but with the imminent cancellation of Centaur and diminishing support for NASA's nuclear propulsion program (which he thought held the key to distant travels in space), he sensed "a new kind of disillusion" that eroded confidence in what he referred to as man's "extraterrestrial imperative."[93]

Ironically, although Centaur's close identification with the Apollo program had initially saved it from cancellation, the successful conclusion of the Surveyor program increased its vulnerability. Centaur had won funding during a time when the urgency of the space race with Soviet Union had provided the NASA with ample funding for facilities and personnel to support the development of a radical technology. Centaur had proven not only the feasibility of liquid hydrogen, but also the viability of the concept of government partnership with industry in developing the space agency's technical competence. However, declining budgets for the space program and lack of unity among scientists over long-term planning threw the planetary program temporarily into limbo.[94] With no future orders for Atlas-Centaur, General Dynamics prepared to shut down its factory. On 6 February 1968, W. L. Gorton, Division Vice-President and General Manager of the Pratt & Whitney Development Center, informed Abe Silverstein at Lewis Research Center that within months, they would complete delivery of all RL10 engines on order. Key RL10 engine personnel would have to be reassigned and production space turned over to other projects. He lamented that the nation was about to lose a precious technical capability in liquid hydrogen. "Although the special manufacturing and testing facilities can be retained and stored in whatever fashion is most practical, there is no way to mothball the know-how, the production team, and the factory space," he wrote. "Previous experience with other engines has demonstrated that once the production team is dispersed, it is most difficult and costly to assemble and train a new team with the required expertise."[95]

---

[93] Quoted by Marsha Freeman, *How We Got to the Moon*, 314.

[94] Robert S. Kraemer, *Beyond the Moon: A Golden Age of Planetary Exploration: 1971–1978* (Washington: Smithsonian Institution Press, 2000), xiv.

[95] W. L. Gorton to Abe Silverstein, 6 February 1968, Box 59, Old RL10 Records (Goette files), DEB Vault, NASA Glenn.

A Changing Vision of Space Travel

Had the Centaur rocket become superfluous? The Apollo years in the 1960s had focused single-mindedly on landing Americans on the lunar surface before their Soviet counterparts. In an era of nearly limitless funds for Cold War projects, the notion of a reusable rocket was not given a high priority.[96] However, once the Moon missions were successfully completed, simply discarding rocket stages that cost millions of dollars to manufacture seemed extravagant. Faced with tightening budgets, NASA leaders decided that seemingly more economical ferrying systems and reusability were the most important goals for the future. Apollo 11 astronaut Michael Collins recalled that in the late 1960s, reusability "implied a technological maturity, a feeling that space was here to stay, that launching ships into Kennedy's 'new ocean' would become routine." Collins concluded that for an aeronautical engineer, "the path to reusability was glorious."[97]

As the 1960s came to a close, NASA began to officially investigate designs for a reusable space vehicle called the Space Transportation System (STS) or Space Shuttle.[98] In 1969, NASA awarded contracts for the first phase of the project. General Dynamics, Lockheed, McDonnell Douglas, and North American Rockwell each received $500,000 to conduct feasibility studies.[99] By September, NASA had pooled these recommendations together and appraised the possible designs. The approved plan was to develop a "reusable-rocket transportation system" that was capable of "routine access to space."[100] When the next phase began in July 1970, reusability was the primary goal, along with a piloted orbiter capable of at least one hundred missions. In early 1972, after a meeting with NASA Administrator James Fletcher and his deputy, George Low, President Richard Nixon officially announced the plan to dramatically change the American initiative in space.[101] The main rationale for the Shuttle was the presumed low cost of a vehicle that could be used many times compared to that of a launch vehicle that was used only once.

---

[96] Stuart W. Leslie, *The Cold War and American Science: The Military-Industrial-Academic Complex at MIT and Stanford* (New York: Columbia University Press, 1993).

[97] Michael Collins, *Liftoff: The Story of America's Adventure in Space* (New York: Grove Press, 1988), 202.

[98] T. A. Heppenheimer, *The Space Shuttle Decision: NASA's Search for a Reusable Space Vehicle* (Washington, DC: NASA, 1999).

[99] David Baker, "Evolution of the Space Shuttle, Part I," *Spaceflight* 15 (June 1973): 202–210.

[100] Grey, *Enterprise*, 59. Space Task group, "The Post-Apollo Space Program: Directions for the Future," September 1969, as found in *Exploring the Unknown*, vol. I, 525–543.

[101] Henry C. Dethloff, "The Space Shuttle's First Flight: STS-1," in *From Engineering Science to Big Science: The NACA and NASA Collier Trophy Winners*, ed. Pamela E. Mack (Washington, DC: Government Printing Office, 1998), 286. See also White House Press Secretary, "The White House, Statement by the President," 5 January 1972, Richard M. Nixon Presidential Files, NASA Historical Reference Collection, NASA Headquarters, Washington, DC, as found in "Nixon Approves the Space Shuttle," Roger D. Launius, *NASA: A History of the U.S. Civil Space Program* (Malabar, Florida: Krieger Publishing Company, 1994), 232.

What NASA did not count on was the exorbitant operational costs of sending people into space, as well as the extremely long time it took to turn the Shuttle around.[102]

As plans for the Shuttle were finalized, General Dynamics and Lewis engineers reconciled themselves to the idea that the Centaur Program would be canceled at some time in the not-too-distant future. Centaur won a reprieve after the transfer of Atlas-Agena missions to Atlas-Centaur in January 1969.[103] Then Atlas-Centaur became the designated launch vehicle for NASA's ambitious planetary program of the 1970s while the nation awaited the Shuttle. In the next decade, a Centaur-launched spacecraft visited every large body in the solar system—the Sun, Earth's Moon, Mercury, Venus, Mars, Jupiter, Saturn, Uranus, Neptune, and beyond.

---

[102] See Introduction by Roger Launius to *To Reach the High Frontier*, 14.

[103] Agena continued as an Air Force single and upper stage until 1985.

Chapter 4

Heavy Lift

"It was one of those rare moments in time. In the brief span of eight years, from 1971 to 1978, Americans launched a fleet of robot spacecraft on paths to the far corners of the solar system . . . undertaking what must surely be the greatest burst of exploration in the history of mankind."

–Robert S. Kraemer, NASA Office of Space Science

The 1970s became one of the most significant decades in the history of space science, a branch of knowledge that received impetus after the launch of Sputnik in October 1957. In 1958, Eisenhower's Science Advisory Committee defined space science in a pamphlet entitled *Introduction to Outer Space* as "knowledge of the Earth, solar system, and the universe."[1] In his book describing NASA's early space program, Homer Newell made clear how space science differed from the work of earlier earthbound astronomers who had trained their telescopes on the heavens. The space program gave scientists new instruments and a new vantage point from which to see the heavens, a vantage point free from many of the optical distortions produced by Earth's atmosphere. Newell called attention to the important post-Sputnik role that rockets and spacecraft played in the pursuit of new knowledge of the solar system. For Newell, space science comprised "those scientific investigations made possible or significantly aided by rockets, satellites, and space probes."[2] NASA's space science program had an enormous impact on the sciences, particularly the disciplines of geophysics, astronomy, physics, and geology. Robert Kraemer, a former director of NASA planetary programs, recalled, "It was one of those rare moments in time. In the brief span of eight years, from 1971 to 1978, Americans launched a fleet of robot spacecraft on paths to the far corners of the solar system . . . undertaking what must surely be the greatest burst of exploration in the history of mankind."[3]

---

[1] Homer Newell, *Beyond the Atmosphere: Early Years of Space Science* (Washington, DC: NASA, 1980), xiii.

[2] Ibid., 11.

[3] Robert S. Kraemer, *Beyond the Moon: A Golden Age of Planetary Exploration: 1971–1978* (Washington: Smithsonian Institution Press, 2000).

During the 1970s, Centaur technology advanced dramatically. The most significant technical innovation during this period was a new Teledyne guidance and control system first tested in 1973. With more advanced avionics, mission planners had more flexibility in the design of missions. The new guidance and control system made possible the development of a revolutionary approach to programming vehicle flightpaths through the winds encountered during liftoff. At the same time, a new rocket propulsion test facility at Lewis's Plum Brook Station in Sandusky, Ohio, permitted more sophisticated ground-testing of Centaur's restart capability. The missions themselves during this period provided opportunities for testing new hardware and software in flight.

Like the spaceflights that sent human beings to the Moon, missions that contributed to new knowledge of the universe had propaganda value in demonstrating America's technical and scientific prowess. Mars and Venus were highly sought-after prizes for whichever nation could be the first to explore them.[4] J. N. James, a JPL engineer during this period, commented, "Although the official U.S. position at that time was that we were not competing with the USSR, my fellow team members and I felt otherwise."[5] The race to the planets put pressure on NASA to take risks. With the benefit of hindsight, Newell wrote that because the Agency was "exposed directly to the outside pressures to match or surpass the Soviet achievement in space, NASA moved more rapidly with the development of observatory-class satellites and the larger deep-space probes than the scientists would have required."[6] These pressures probably contributed to early failures of missions dedicated to the advance of NASA's space science program.

In the 1960s, NASA's long-range plan was to send flyby missions to the closest planets to Earth. These missions would pave the way for more complex orbiting missions, eventually leading to landing on a planet. In 1960, JPL concluded that Centaur would make the ideal upper stage for Mariner but opted for the less powerful Atlas-Agena when Centaur development problems threatened cancellation of the program. The lack of reliable launch vehicles "bedeviled the American space endeavor from the beginning," wrote historians Clinton and Linda Ezell. "While this relationship between launch vehicle and spacecraft was apparent in any space project, it had an especially negative effect on Mariner."[7] With the American planetary program held back by the lack of a sufficiently powerful and reliable upper stage, the Soviet Union launched the first probe to Mars in 1962. Relief over the fact that the Russian probe never

[4] James Schefter, *The Race: The Uncensored Story of How America Beat Russia to the Moon* (New York: Doubleday, 1999); William H. Schauer, *The Politics of Space: A Comparison of the Soviet and American Space Programs* (New York: Holmes & Meier, Publishers, 1976).

[5] J. N. James, "The First Mission to Mars," in *Mars: Past, Present, and Future*, ed. E. Brian Pritchard, vol. 145: *Progress in Astronautics and Aeronautics*, 1992, 30.

[6] Homer Newell, *Beyond the Atmosphere*, 97.

[7] Edward Clinton Ezell and Linda Neuman Ezell, *On Mars: Exploration of the Red Planet, 1958–1978* (Washington: NASA SP-4212, 1984), 25.

reached the planet probably did little to calm NASA's sense of urgency. Mariner 1 attempted a Venus flyby in 1962, but range safety officials had to destroy it when it veered off course. A Mars flyby with Mariner 3 in 1964 went silent 9 hours after launch. The second attempt at missions to each of these planets proved more successful. Mariner 2 became the first spacecraft to encounter another planet when it reached Venus in 1962, and Mariner 4 provided the first close view of the mysterious red planet in 1964.

However, mission specialists in the 1970s needed the heavy lift capability of Centaur. Robert Gray, Director of Unmanned Launch Operations at Kennedy Space Center, captured the essence of this era in two words—"heavy science."[8] The high thrust capability of Centaur, achieved through the use of liquid-hydrogen/liquid-oxygen propellants, made it the upper stage of choice for JPL's scientific missions in the 1970s. As Centaur proved its reliability, JPL personnel formed strong working relationships with their counterparts at General Dynamics and Lewis Research Center. Joe Nieberding attributed this change in perception to the increasing competence of Lewis Research Center Centaur managers. "They recognized that we knew what we were doing," he said. "We really depended on each other to do our jobs."[9] Mission designer Charles Kohlhase of JPL chose Centaur because of the vehicle's "enormous performance capability."[10] Norm Haynes of JPL recalled that in 1971, Atlas-Centaur was "the only way we could get a spacecraft into Mars orbit . . . we started out with a big mass and we started shaving a little in the spacecraft, but we got most of it out of the launch vehicle."[11]

Atlas-Centaur sent very heavy orbiting observatories to circle Earth and record extensive data about the celestial sky. It also sent a variety of spacecraft on trips throughout the solar system.[12] Titan-Centaur, to be discussed in the following chapter, launched six missions to the Sun and the inner and outer planets. Two decades later, a redesigned Titan-Centaur launched the Cassini mission to Saturn. NASA's ambitious probes to the outer reaches of the solar system helped win the prestige race with the Soviet Union. Because the Russians used conventional fuel in their upper stages, their craft were heavier and could not provide the same weight-carrying capacity as Centaur.

---

[8] Interview with Robert Gray by Virginia Dawson, 9 November 1999.

[9] Interview with Joe Nieberding by Virginia Dawson and Mark Bowles, 15 April 1999.

[10] Interview with Charles Kohlhase by Virginia Dawson, 8 June 2000.

[11] Interview with Norm Haynes by Virginia Dawson, 9 June 2000.

[12] The nine planetary launches, together with the six launches of Titan III-Centaur discussed in the following chapter, constitute almost the entire history of Centaur launched probes to uncover the mysteries of our solar system. (The first of the seven Titan III-Centaur launches was a mass simulator payload.) There were two later Atlas-Centaur launches: the Atlas-Centaur IIAS launch of the Solar Heliospheric Observatory (SOHO) for NASA by the Atlas commercial program in December 1995 (AC-121) and the Titan IV-Centaur launch of Cassini for NASA in October 1997 (TC-21). SOHO was put into a LaGrange orbit to study the Sun, and Cassini was a science mission to the outer planets.

In addition to its service to NASA's space science program, Atlas-Centaur became the preferred launch vehicle for the International Telecommunications Satellite Organization (Intelsat). During the 1970s, Atlas-Centaur launched nineteen commercial satellites (with two failures) and the first of seven Fleet Satellite Communications (FLTSATCOM) satellites for the military. These satellites were the spaceborne portion of a global Department of Defense communications system. The decade proved exceptionally busy for the Launch Vehicles Division at Lewis, headed between 1971 and 1974 by Dan Shramo. As the Lewis role in launch vehicles expanded in 1974, Center Director Bruce Lundin created a Launch Vehicles Directorate under Andrew Stofan. Henry Slone became manager of Atlas-Centaur, with Paul Winslow replacing Stofan as project manager of Titan-Centaur.

A Growing Expertise

A massive downsizing of Lewis Research Center in the early 1970s could not dampen the enthusiasm of the members of the Launch Vehicles Division for their bird. No longer Abe's baby, Centaur belonged to the new generation of men and women who had experienced the tribulations of developing Centaur. The young engineers whom NASA had recruited in the early 1960s were just reaching their stride. Not part of the laboratory's research tradition, they had matured with the Centaur program, shaped by the constant pressure of a launch date dictated by the position of Earth in relation to the other planets. From their common focus and physical isolation from the rest of the laboratory evolved a distinctive launch vehicles culture within Lewis. Joe Nieberding, a graduate of local John Carroll University with a major in physics, had spent his entire life in Cleveland before he started work at NASA Lewis in the early 1960s. Within his first month in the Launch Vehicles Division, he had seen both the Atlantic and Pacific Oceans for the first time.[13]

People in the Launch Vehicles Division grew used to shuttling back and forth across the country. They attended meetings, scrutinized the fabrication of Centaur tanks in San Diego, monitored tests at Sycamore Canyon and Point Loma, negotiated with the payload specialists at JPL, set up shake tests at the Lewis Plum Brook facility, and flew down to Cape Canaveral to prepare for launches. Some of the older "research men" at Lewis resented the privileges and high salaries of the new recruits in the Launch Vehicles Division, but they wanted nothing to do with the rough-and-tumble world of Centaur. The Development Engineering Building (DEB) completed in the mid-1960s provided office space for several hundred engineers outside the laboratory's main gate. The location of the Launch Vehicles Division in the DEB emphasized its separation from the research side of the laboratory.

Because of the importance of the interface between the payload and the launch vehicle, Centaur engineers became thoroughly familiar with the science of the missions. Most often,

---

[13] Interview with Joe Nieberding by Virginia Dawson and Mark Bowles, 15 April 1999.

they worked with mission planners at JPL in Pasadena, California. They also managed missions for Goddard Space Flight Center, Ames Research Center, and Marshall Space Flight Center during this period. Lewis and General Dynamics launch vehicle engineers were invited to the key science meetings so that they could better understand the Centaur payload requirements. Andrew Stofan said, "We attended all of their major reviews, so we were part of each science spacecraft and, of course, that meant we had to have a lot of people in a lot of places, because we were part of every science project being put on the launch vehicle." Scientists reciprocated by taking a hands-on approach to Centaur's technology. "They don't sit back up in the ivory tower," Tom Shaw of JPL remarked. "They're out there kicking tires regularly."[14]

Mission planning began two or three years before the actual launch. The Centaur team liked to joke that, almost without exception, the payload always started out heavier than the Centaur could lift. Mission interface, handled by a group working under Edwin Muckley of Lewis, tackled questions such as how much can the spacecraft weigh? How much power does it need from the rocket? What is the target? What is the mission? How fast does the rocket have to go? How much heat can the spacecraft take? How much vibration can it stand when the rocket lifts off?

Another important way in which the Lewis launch vehicle team influenced a mission was through trajectory analysis. The team began its work by defining the key variables. Some of these were constant among all missions. For example, Earth always rotated and revolved around the Sun with near-constant speeds because of its almost circular orbit. Other variables were mission-specific. Different planets had vastly different trajectories. Once the destination was known, other variables could be defined, such as the weight of Centaur, which changed with the differing fuel requirements, and the weight of the payload, which depended on the type of science instruments to be flown.[15]

Because Centaur was launched from a revolving, rotating platform to a revolving, rotating target, trajectory calculations changed depending upon the time of launch. A two-day launch opportunity required a trajectory for each possible minute of liftoff. Members of the Centaur trajectory team at Lewis pioneered many of the advanced trajectory calculation techniques needed to handle planetary mission design.

Lewis Omer (Frank) Spurlock, working under Fred Teren in the Lewis Performance Trajectory Group, became an expert in calculating trajectories. Spurlock had graduated from the University of New Mexico and won a Woodrow Wilson fellowship to Western Reserve University (later renamed Case Western Reserve University) to study physics. Just twenty-one when he started at Lewis in 1961, he traded graduate work for the chance to contribute to the space program at a time when knowledge of how to calculate the trajectory of a spacecraft was in its infancy.[16]

[14] Interview with Tom Shaw by Virginia Dawson, 10 November 1999.

[15] Richard T. Mittauer, "Mariner Mars 1971 Launches," 30 April 1971, Box AC-15 to AC-24, Division Atlas/Centaur Project Office, NASA GRC Records.

[16] Oran Nicks, Far Travelers, 62.

Designing missions for expendable launch vehicles involves generating a mathematical model that takes into account launch vehicle performance capabilities and trajectory characteristics over a range of mission requirements and launch periods. For this task, Spurlock started out with an IBM computer with only 8 kilobytes of memory. He used an analog computer for the atmospheric portion of the trajectory. He then manually completed the upper stage calculations with the tiny digital computer. Lewis's acquisition of an IBM 7090 computer with 32 kilobytes of memory greatly increased Spurlock's computational capability. At first, Spurlock turned the trajectory calculations for each mission over to Harry Dempster at General Dynamics to work up into particular payload specifications. Then Spurlock developed an optimization computer code known as DUKSUP. (Pronounced "duck soup," it was not an acronym but a reference to the ease with which it could be applied.) DUKSUP provided an alternate, very accurate means of determining a mission trajectory. It allowed mission planners to model various mission scenarios. The analytic techniques used in DUKSUP depended on an accurate vehicle model and the power of the calculus of variations. Early versions of the code were used on Surveyor, Mariners 6 through 10, Pioneers 10 and 11, Viking, Voyager, Helios, Pioneer Venus, the High-Energy Astronomy Observatory (HEAO), FLTSATCOM, INTELSAT, the Orbiting Astronomical Observatory (OAO), and Applications Technology Satellites (ATS). DUKSUP also gave government engineers an analytic tool to evaluate contractor data.[17]

After Surveyor

The first assignment for Centaur after Surveyor was to launch two Department of Defense satellites needed for meteorological and space environment tests. Although Atlas-Agena had launched the previous satellites in this series, Centaur was chosen for ATS 4 and 5 because they were too heavy for Agena to lift.[18] Some key modifications were required to adapt Centaur technology for this new mission. Previous Centaurs were capable of a 25-minute coast period, but the ATS needed a 60-minute coast. For attitude control and propellant settling during the long coast period, engineers increased the supply of hydrogen peroxide. To compensate for the extreme cold Centaur would encounter during the coast period, they developed better thermal control sensors. However, the most significant change was to design a new satellite adapter to replace the one that had attached Surveyor to Centaur. The modifications

[17] L. R. Balkanyi and O. F. Spurlock, "DUKSUP: A High Thrust Trajectory Optimization Code," AIAA 93-1127, Irvine, CA, 16–19 February 1993; and an interview with Frank Spurlock by Virginia Dawson, 6 April 1999. Much of the documentation of its early applications has been lost.

[18] Donald H. Martin, *Communication Satellites*, 4th edition (California: The Aerospace Press, 2000), 18. For further information on ATS, see R. H. Pickard, "The Applications Technology Satellite," in *Proceedings of the 16th International Astronomical Congress* (1965), vol. 4: *Meteorological and Communications Satellites* (1966); and Paul J. McCeney, "Applications Technology Satellite Program," *Acta Astronautica* 5 (March–April 1978): 299–325.

begun on 8 April were ready by 24 July 1968. In all, there were thirty significant changes. The Centaur Project Office at Lewis Research Center compiled the following summary of technical modifications to Centaur in preparation for post-Surveyor missions.[19]

Centaur Technical Modifications After Surveyor Missions Were Completed		
System	Change from Surveyor	Reason for Change
Mechanical Systems, Airborne, Peroxide System	Provide a dual-bottle H_2O_2 supply system and the necessary support structure.	Provide an additional H_2O_2 supply to cover the extended coast period.
Mechanical Systems, Airborne, Peroxide System	Provide for improved thermal control of the H_2O_2 bottles and the boost pump overspeed sensors.	Required to compensate for the possible temperature extremes resulting from the extended coast period.
Mechanical Systems, Airborne, Peroxide System	Requalify the 3-pound thrust H_2O_2 motors for 65-minute operation.	Required to meet the extended coast-period requirement.
Mechanical Systems, Airborne, Pneumatics	Redesign the vent valve friction devices and controller bellows.	General design improvement.
Mechanical Systems, Airborne, Pneumatics	Requalify the pneumatic regulators to −45°F.	Required as a result of possible lower temperature environment during the extended coast period.
Mechanical Systems, Airborne, Propellants	Propellant utilization electronics package to be built and tested by GD/CC instead of subcontractor.	Difficulties with subcontractor.
Mechanical Systems, Airborne, Structural, Nose Fairings	Add two access ports on the + and − y axes.	Provide access to the ATS electrical connectors.

[19] "Presentation of Launch Vehicle System in Support of the ATS-D Mission," 16 April 1968, Glenn Research Center, DEB Archives, Box AC-15 to AC-24, Division Atlas/Centaur Project Office.

System	Change from Surveyor	Reason for Change
Mechanical Systems, Airborne, Structural, Nose Fairings	Add two electrical disconnect support arms.	Support spacecraft safe/arm and power supply cabling and withdraw the disconnect cards from the spacecraft during nose-fairing jettison.
Mechanical Systems, Airborne, Structural, Nose Fairings	Add spacecraft T&C antenna ramps and rub surfaces.	Prevent the spacecraft antennas from 1) being abraded and 2) hanging up on the fairing longerons during nose-fairing jettison.
Mechanical Systems, Airborne, Structural, Nose Fairings	Delete fiberglass air conditioning ducts from the conical fairing and add removable metal ducts.	Make the ATS duct configuration common with that required for Mariner '69 and allow removal of the duct for cleaning.
Mechanical Systems, Airborne, Structural, Nose Fairings	Relief cuts made on thermal bulkhead at nose-fairing split line.	Provide sufficient clearance between the thermal bulkhead and transition adapter during nose-fairing jettison.
Mechanical Systems, Airborne, Structural, Nose Fairings	Delete Surveyor-peculiar TV lights and add ATS TV target lights.	Meet ATS requirements to check out spacecraft cameras.
Mechanical Systems, Airborne, Structural, Payload Adapter	Delete Surveyor forward payload adapter and add an ATS-peculiar transition adapter.	Required to join the ATS (HAC) adapter with the Centaur field joint.
Mechanical Systems, Airborne, Structural, Payload Adapter	Modify the electrical interface island.	Accommodate the ATS destructor leads and the payload separation and instrumentation connectors.
Mechanical Systems, Payload Adapter	Install new air conditioning ducts in the transition adapter.	Conform to the design of the new adapter and satisfy the ATS thermal requirements.

System	Change from Surveyor	Reason for Change
Electrical System, Airborne	Delete Surveyor-peculiar harnessing.	Not required to support ATS mission.
Electrical System, Airborne	Provide a pyro harness from the pyro relay package to the adapter interface island.	Meet ATS requirement for Centaur to provide spacecraft separation electrical power.
Electrical System, Airborne	Add harnesses between Centaur umbilical and 1) the spacecraft and 2) TV lights on the nose fairing.	Meet ATS requirements for 1) spacecraft safe/arm and power supply circuits and 2) nose-fairing TV lights.
Electrical System, Airborne	Install a larger capacity (150 ampere-hour) battery with a voltage monitor resistor.	Meet need for additional Centaur electrical power required because of the extended coast and retromaneuver periods.
Electrical System, Instrumentation and Telemetry Systems	Delete Surveyor-peculiar instrumentation and add instrumentation for additional Centaur data.	Primarily provide for retrieval of data associated with the extended coast. Also provide for analysis of the spacecraft adapter interface vibrations during flight.
Electrical System, Instrumentation and Telemetry Systems	Add a second telepak.	Provide for increased number of measurements.
Electrical System, Instrumentation and Telemetry Systems	Provide for two ATS acceleration measurements.	Meet ATS requirement.
Guidance and Autopilot, Atlas	Remove pitch program from Atlas programmer and place in Centaur computer.	Effect system simplification.

System	Change from Surveyor	Reason for Change
Guidance and Autopilot, Atlas	Revise the programmer sequence to delay the nose-fairing jettison event.	Reduce the aerodynamic heating input to the spacecraft and establish commonality with AC-19 and AC-20.
Guidance and Autopilot, Centaur	Redesign rate gyro package stability margin.	Provide increased vehicle to provide additional levels and switching for rate and position gains.
Guidance and Autopilot, Centaur	Provide nonlinear position gains during the reorientation interval.	Enable the accomplishment of a large angle reorientation maneuver in a limited time interval and with a high degree of accuracy.
Guidance and Autopilot, Centaur	Relocate the Spin Motor Rotation Detector external to the rate gyro package.	Provide room for the additional components required in the rate gyro package to effect the above changes.
Guidance and Autopilot, Centaur	Delete Surveyor-peculiar functions from the programmer and add the ATS discretes.	Required primarily to implement the reorientation and retromaneuver sequences.
Guidance and Autopilot, Centaur	Revise the timer logic to reference spacecraft separation to the Main Engine Cut Off event.	Enable more rapid retromaneuver due to elimination of fixed time delays for burn-time dispersions.
Guidance and Autopilot, Centaur	Revise programmer switching to effect a longer prestart sequence.	Compensate for anticipated higher temperatures resulting from long coast.

Since satellite technology was still in the experimental stage, the main goal for ATS 4 and 5 (AC-17 and -18) was to investigate how to maintain satellite stability while in orbit.[20] Both launches were disappointing. In August 1968, Centaur separated from the Atlas and coasted successfully for just over an hour, but its engines failed to restart. A year later, Centaur successfully placed the satellite into transfer orbit after a 25-minute coast, then developed an unplanned-for spin that compromised its communications experiments.[21] Because of its poor performance, Centaur was abandoned for the final ATS launch in 1974 in favor of the Titan IIIC vehicle.

After this inauspicious beginning, the Centaur launch of a revolutionary, very heavy Orbiting Astronomical Observatory (OAO), managed by Goddard Space Flight Center, was more successful.[22] The great weight of OAO was an unavoidable attribute because it required extremely sophisticated optical instrumentation. Considerable weight was also taken up by basic equipment such as power supplies, temperature control, and tracking and telemetering equipment. The first OAO, launched by Atlas-Agena D in 1966, had failed.[23]

OAO-2, launched in December 1968, weighed 4,450 pounds.[24] Its main objective was to make unprecedented, precise telescopic observations 480 miles above Earth, where the atmosphere produced fewer optical distortions. Because the payload was too large for the Atlas-Centaur nose fairing, engineers used the Atlas-Agena fairing and connected the OAO to the front of Centaur with the Agena adapter.[25] Four days after a flawless Atlas-Centaur launch, the OAO onboard experiments immediately began showing the presence of star fields, returning thousands of unique images and mapping the northern skies.[26]

An even heavier and more complex scientific satellite, OAO-3, was launched in November 1970 but never reached orbit. When the Centaur computer attempted to jettison the nose fairing

---

[20] Daniel R. Glover, "NASA Experimental Communications Satellites, 1958–1995," in *Beyond the Ionosphere: Fifty Years of Satellite Communication* (Washington, DC: NASA, 1997), 56–60.

[21] Roy K. Hackbarth, "Atlas-Centaur AC-18 Performance Evaluation," NASA TM X-2383, Box AC-15 to AC-24, Division Atlas/Centaur Project Office, NASA GRC Records.

[22] "AC-16 Centaur Flight Evaluation Report," Glenn Research Center, DEB Archives, Box AC-15 to AC-24, Division Atlas/Centaur Project Office. See also Mike Reynolds, "The Orbiting Astronomical Observatories," in *USA in Space*, eds. Frank N. Magill and Russell R. Tobias (Pasadena, CA: Salem Press, Inc., 1996), 471; and Homer Newell, *Beyond the Atmosphere*, 145.

[23] "The Orbiting Astronomical Observatory," *Sky and Telescope* (December 1962): 339–340.

[24] "Atlas/Centaur-16 Orbiting Astronomical Observatory-2 Final Field Report," 17 October 1969, Box AC-15 to AC-24, Division Atlas/Centaur Project Office, NASA GRC Records.

[25] "AC-16 Centaur Flight Evaluation Report," Glenn Research Center, DEB Archives, Box AC-15 to AC-24, Division Atlas/Centaur Project Office. See also G. R. Richards and Joel W. Powell, "The Centaur Vehicle," *British Interplanetary Society* 42 (1 March 1989): 108.

[26] Arthur Code quotation found in "OAO-B Launch," General Release, Box AC-15 to AC-24, Division Atlas/Centaur Project Office, NASA GRC Records. Arthur D. Code, ed., *The Scientific Results from the Orbiting Astronomical Observatory* (Washington, DC: NASA SP-310, 1972).

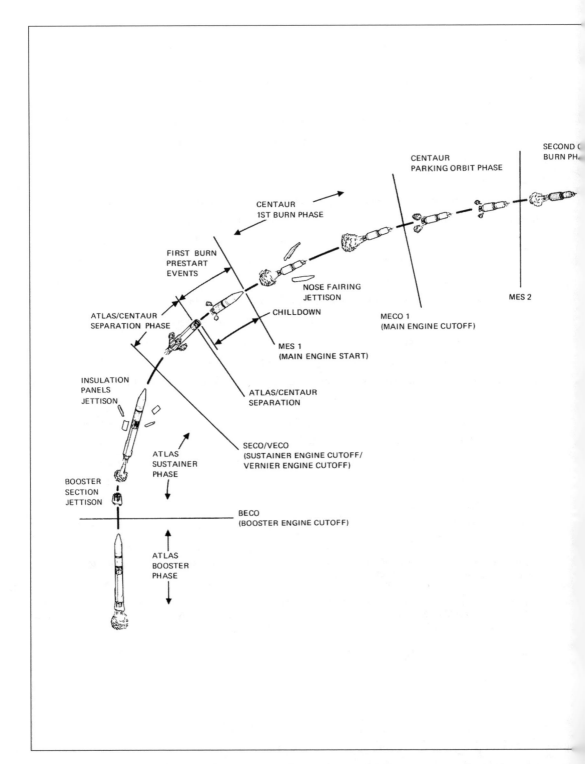

Atlas-Centaur parking orbit mission delivering a spacecraft to synchronous apogee transfer. (Courtesy of Lockheed Martin. See
Centaur: Mission Planners Guide, *File 010216, NASA Historical Reference Collection.)*

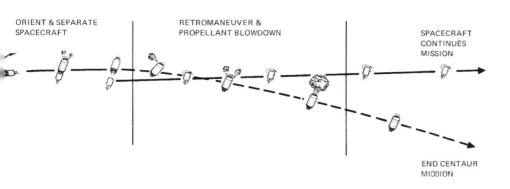

ORIENT & SEPARATE
SPACECRAFT

RETROMANEUVER &
PROPELLANT BLOWDOWN

SPACECRAFT
CONTINUES
MISSION

END CENTAUR
MISSION

Atlas-Centaur Launches of Scientific Satellites				
Date	Mission	Vehicle	Payload Weight	Result
10 August 1968	ATS-4	AC-17	130 lb*	Centaur failed
7 December 1968	OAO-2	AC-16	4,447 lb	Success
12 August 1969	ATS-5	AC-18	130 lb*	Success
30 November 1970	OAO-3	AC-21	4,698 lb	Centaur failed
21 August 1972	OAO-4	AC-22	4,914 lb	Success
12 August 1977	HEAO-1	AC-45	5,549 lb	Success
13 November 1978	HEAO-2	AC-52	6,866 lb	Success
20 September 1979	HEAO-3	AC-53	6,321 lb	Success

*The actual weights carried by Centaur to transfer orbit were approximately double these numbers.

Atlas-Centaur Launches of Planetary Science Missions				
Date	Mission	Vehicle	Payload Weight	Result
24 February 1969	Mariner 6	AC-20	850 lb	Mars flyby
27 March 1969	Mariner 7	AC-19	848 lb	Mars flyby
9 May 1971	Mariner 8	AC-24	2,192 lb	Mars orbiter; Centaur failed
30 May 1971	Mariner 9	AC-23	2,201 lb	Mars orbiter
2 March 1972	Pioneer 10	AC-27	568 lb	Jupiter and the outer planets
5 April 1973	Pioneer 11	AC-30	562 lb	Jupiter and the outer planets
3 November 1973	Mariner 10	AC-34	1,108 lb	Venus and Mercury flyby
20 May 1978	Pioneer Venus	AC-50	1,222 lb	Venus orbiter
8 August 1978	Pioneer Venus	AC-51	1,948 lb	Venus Multiprobe mission

around the payload, one of the sixteen explosive bolts failed to release, and the rocket and satellite plunged to Earth together.[27] Because of this failure, the Centaur fairing was significantly modified. Full-scale jettison tests, under the direction of William Prati, were conducted in the Altitude Wind Tunnel at Lewis until the team had confidence in the new design.

No unforeseen glitches occurred when Atlas-Centaur (AC-22) lifted Copernicus, the fourth and final spacecraft in the OAO series, into orbit on 21 August 1972. OAO-4 carried telescopes designed by scientists at Princeton and University College of London. At 4,900 pounds, it was NASA's heaviest satellite, and it continued to send data back to Earth for nine years.[28] Despite the two failures of the OAO program, the second and fourth satellites "were widely judged to be technical and scientific successes, and that bolstered NASA's confidence that a space telescope much larger than the OAO series could be built."[29] They marked the beginning of above-the-atmosphere observatories that now include the Hubble telescope.[30]

First Centaur Interplanetary Missions: Mariners 6 Through 9

Although the Orbiting Astronomical Observatories had provided unprecedented observations of the sky, this achievement paled in comparison with sending a spacecraft to the vicinity of a planet. Earlier Mariner spacecraft had avoided using Centaur, but the next generation needed the additional weight that only Centaur could carry. By using Centaur, Mariners 6 through 9 carried heavy atmospheric and topographical instruments needed for gathering data in preparation for a future Mars landing.[31] Because failures were so common in the early space program, NASA built two identical Mariner spacecraft for each mission and launched them in pairs.[32]

---

[27] "AC-21 Flight Evaluation Report," Box AC-15 to AC-24, Division Atlas/Centaur Project Office, NASA GRC Records.

[28] Raymond N. Watts, Jr., "An Astronomy Satellite Named Copernicus," *Sky and Telescope* (October 1972): 231–232, 235. Lyman Spitzer, Jr., *Searching Between the Stars* (New Haven: Yale University Press, 1982).

[29] Smith, *The Space Telescope*, 44.

[30] Although no comprehensive history of the OAO exists, several other articles from *Sky and Telescope* document some of its achievements. For further information, see "Observing a Comet from Space," *Sky and Telescope* 39 (March 1970): 143; "The Orbiting Astronomical Observatory," *Sky and Telescope* 24 (December 1962): 339–340; Watts, "More About the OAO's," *Sky and Telescope* 28 (August 1964): 78–79; and "Orbiting Astronomical Observatory," *Sky and Telescope* 28 (August 1964): 78–79.

[31] "Project Mariner Mars 1969," 14 February 1969, Box AC-15 to AC-24, Division Atlas/Centaur Project Office, NASA GRC Records.

[32] Mariner 6 was 271 pounds heavier than Mariner 4 and 399 pounds heavier than Mariner 2. For further information on Mariners 6 and 7, see James Wilson, *Two Over Mars: Mariner VI and Mariner VII, February–August 1969* (Washington, DC: NASA EP-90 Government Printing Office, 1971); Clark R. Chapman, *Planets of Rock and Ice* (New York: Charles Scribner's Sons, 1982); Scientific and Technical Information Division, Office of Technology Utilization, *Mariner Mars 1969: A Preliminary Report* (Springfield, VA: NASA SP-225 Clearinghouse, Department of Commerce, 1969).

Mariner 6 (AC-20) was nearly marred by a catastrophic accident ten days before its scheduled liftoff. Atlas-Centaur and its Mariner payload were sitting on the launch pad at the Cape
when Atlas suddenly began to lose pressurization. A faulty relay switch had caused one of the
main valves to open. As the Atlas began to slowly crumple, two General Dynamics mechanics
ran under the twelve-story assembly with its deflating booster to seal off the pressure manually
by closing a valve. The *Lewis News* reported the heroic story:

> Ignoring the shrill scream of the evacuation alarm, Billy McClure hastily removed the
> liquid oxygen prevalve locking bolts to the Atlas vehicle on Pad 36A at the Cape. As
> he did this, he heard the sound of metal buckling, and looking upward he saw the Atlas
> begin to lean under the weight of the Centaur and Mariner spacecraft as the pressure
> leaked out of the vehicle. At the same time Charles Beverlin had run under the Atlas
> and squirmed up into the Atlas thrust section to close the sustainer engine prevalve.
> Seconds later both mechanics had cleared the area, and the vehicle began to right itself
> as the tank pressure built up inside.[33]

Without their quick action, the mission would have been lost. They received a rare award
given by NASA, the Exceptional Bravery Medal.[34] Liftoff in February 1969 went as scheduled
with a replacement Atlas. It flew by Mars in July at an altitude of 2,170 miles.[35] Total flight
time was 156 days.

Mariner 7 (AC-19), launched in March, had almost identical mission objectives to those of
Mariner 6, except that it was to fly by the Martian southern polar cap.[36] The relative locations of
Earth at launch and Mars at encounter allowed Mariner 7 to use a direct-ascent trajectory.
Together, the two spacecraft photographed twenty times more surface area of the planet than
Mariner 4. This included photographing the surface of the planet, studying the Martian atmosphere,
and also performing spectrometer studies to determine whether the conditions for life were
present on Mars. The atmospheric and surface data gathered by the spacecraft led scientists to
conclude that life probably did not exist on Mars unless it was microbial.[37] In discussing the
significance of these missions, Stuart Collins wrote, "Mariners 6 and 7, launched by the more

---

[33] "Mechanics Risk Lives on Launch Pad," *Lewis News*, Box Awards Folders Through 1979, Folder Exceptional Bravery
(NASA), NASA GRC Records.

[34] Abe Silverstein to NASA Headquarters, 15 April 1969, Box Awards Folders Through 1979, Folder Exceptional Bravery
(NASA), NASA GRC Records.

[35] "Atlas-Centaur 20 Mariner Mars-6 Final Field Report," 19 January 1970, Box AC-15 to AC-24, Division Atlas/Centaur
Project Office, NASA GRC Records.

[36] "Atlas-Centaur-19 Mariner Mars-7 Final Field Report," 19 January 1970, Box AC-15 to AC-24, Division Atlas/Centaur
Project Office, NASA GRC Records.

[37] Henry W. Norris, "Mariner 6 and 7," in *USA in Space*, 366–369.

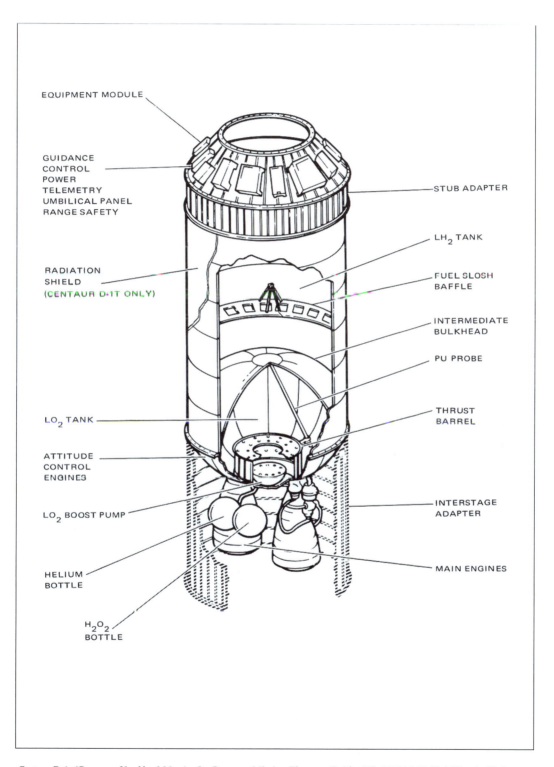

EQUIPMENT MODULE

GUIDANCE
CONTROL
POWER
TELEMETRY
UMBILICAL PANEL
RANGE SAFETY

STUB ADAPTER

LH$_2$ TANK

FUEL SLOSH
BAFFLE

RADIATION
SHIELD
(CENTAUR D-1T ONLY)

INTERMEDIATE
BULKHEAD

PU PROBE

LO$_2$ TANK

THRUST
BARREL

ATTITUDE
CONTROL
ENGINES

INTERSTAGE
ADAPTER

LO$_2$ BOOST PUMP

MAIN ENGINES

HELIUM
BOTTLE

H$_2$O$_2$
BOTTLE

Centaur D-1. (Courtesy of Lockheed Martin. See Centaur: Mission Planners Guide, *File 010216, NASA Historical Reference Collection.)*

powerful Atlas-Centaur rocket, were to greatly increase and improve the observations made by Mariner 4 and by Earth-based investigators. Thus scientists and engineers began to design a more ambitious exploratory mission and to build larger, more sophisticated spacecraft to complete this mission."[38]

What distinguished the two additional Mariner missions launched in 1971 from earlier trips to Mars was that each spacecraft was intended to orbit Mars for a period of 90 days, not merely perform a flyby of the planet. The weight-carrying capability of Centaur was again critical to the success of these missions. Each of the new Mariners weighed 2,197 pounds, more than twice the weight of Mariners 6 and 7. When Mariner 8 (AC-24) lifted off on 9 May 1971, everything went as planned until moments after Centaur and Mariner separated from Atlas, 255 seconds into the flight. Robert Kraemer remembers that, as he sat next to Dan Schneiderman at the consoles in the JPL Mission Control Center at the Cape, he gave the thumbs-up sign to Schneiderman as soon as the Centaur engines ignited. Turning back to the console, he noticed that the pitch recorder was showing unstable oscillations about the Centaur pitch axis.[39] Then Centaur began tumbling end over end and plunged, with the spacecraft, into the Atlantic Ocean 350 miles north of Puerto Rico.

Immediately, there was speculation that the guidance system was the culprit. Bill O'Neil, JPL navigation chief for the spacecraft, quickly discounted the possibility that the entire guidance system was at fault. He suspected that the problem was merely a defective autopilot since it had occurred just at the point when the system was supposed to turn on.[40] Investigation occurred at breakneck speed under the leadership of George Low, NASA's Deputy Administrator. Time was of the essence because if the next Mariner did not launch within a month, it would be two years before the next opportunity. Ed Ziemba, an engineer on the Lewis Centaur team, and Chet Norris, Kennedy Space Center's lead guidance engineer on Centaur, determined that O'Neil was right. An integrated circuit in the pitch rate gyro preamplifier had failed, not the entire guidance and control system. This tiny, $5 integrated circuit, located in the autopilot system, had caused the failure of the $70-million rocket.[41]

With the malfunction corrected, Mariner 9 (AC-23) lifted off twenty-one days after the Mariner 8 loss. Centaur operated flawlessly. Four months later, it began a Martian orbit—the first time a spacecraft had orbited another planet. Engineers were able to assign some of the tasks orig-

[38] Stewart A. Collins, *The Mariner 6 and 7 Pictures of Mars* (Washington, DC: NASA SP-263 Government Printing Office, 1971), 6.

[39] Robert S. Kraemer, *Beyond the Moon: A Golden Age of Planetary Exploration 1971–1978* (Washington: Smithsonian Institution Press, 2000), 50.

[40] Interview with William O'Neil by Virginia Dawson, 9 June 2000.

[41] Interview with Ed Ziemba by Virginia Dawson, 19 May 1999. See also "Atlas-Centaur 24 Mariner Mars-8 Final Field Report," 26 February 1973, Box AC-15 to AC-24, Division Atlas/Centaur Project Office, NASA GRC Records.

Assembly of OAO shroud in the Altitude Wind Tunnel. (GRC C1965-1458)

A technician makes final checks on Mariner 9, which was launched by Atlas-Centaur 23 to Mars on 30 May 1971. (NASA KSC_71P_0366)

inally scheduled for the ill-fated Mariner 8, such as mapping the Martian surface, to Mariner 9. However, during orbit around Mars, the spacecraft encountered a massive dust storm in the Noachis region, and all of its high-resolution cameras simply returned pictures of dust.[42] A month later, after the storm cleared, Mariner returned striking images of Martian volcanoes, canyons, polar caps, and the satellites Phobos and Deimos.[43] Journalist Clyde Curry Smith reported that Mariner "revolutionized the understanding of a very old 'New Mars.'"[44]

Despite the successful investigation into what had gone wrong with Mariner 8 and the brilliant achievements of Mariner 9, Congress remained extremely critical of NASA's management of

---

[42] Interview with William O'Neil by Virginia Dawson, 9 June 2000.

[43] For further information on Mariners 8 and 9, see *Exploring Space with a Camera*, ed. Edgar M. Cortright (Washington, DC: NASA SP-168 Government Printing Office, 1968); William K. Hartman and Odell Raper, *The New Mars: The Discoveries of Mariner 9* (Washington, DC: NASA SP-337 Government Printing Office, 1974); Patrick Moore and Charles A. Cross, *Mars* (New York: Crown Publishers, 1973); and NASA Scientific and Technical Information Office, *Mars as Viewed by Mariner 9: A Pictorial Presentation by the Mariner 9 Television Team and the Planetology Program Principal Investigators* (Washington, DC: NASA SP-329 Government Printing Office, 1976).

[44] Clyde Curry Smith, "Mariner 8 & 9," in *USA in Space*, 370–374.

Centaur. In addition to the loss of Mariner 8, Centaur had caused the loss of an ATS in August 1968 and an OAO in November 1970. In June 1971, during congressional hearings into these failures, NASA Deputy Administrator George Low admitted that while every effort was made to thoroughly check Centaur systems before launch, it was impossible to guarantee launch success, no matter what level of caution. These losses, he said, served as "harsh reminders that space exploration is still a very difficult business, and, in spite of dedicated personnel and intense attention to details, failures must be expected occasionally."[45]

The hearing reminded the Centaur team that they were still under the watchful eye of the House Committee on Science and Astronautics and needed to improve Centaur performance. Members not only were critical of the three recent failures, but also questioned the overall success rate of Atlas-Centaur. Congressman Joseph Karth stated that an 80-percent success rate was "totally unacceptable for a vehicle on which we have spent about half a billion dollars or more."[46] He thought NASA should consider a substitute launch vehicle. John Naugle, Associate Administrator for Space Science and Applications, responded, "I think it would be an unmitigated disaster too if we stopped the ongoing program."[47] Karth warned NASA that although the Committee had strongly supported the launch vehicle and the planetary space program, future support was not a foregone conclusion. He said, "Unless the reliability numbers change, Dr. Naugle, I don't think we are going to be as kind in the future."[48] Despite this scrutiny, Atlas-Centaur continued to serve as the launch vehicle of choice for NASA's program of planetary exploration. Indeed, the power of the Centaur stage made more elaborate missions possible.

To Jupiter and the Solar System: Pioneer 10

While interplanetary explorations of the early 1970s were limited to the nearest neighbors on either side of Earth—Venus and Mars—scientists eagerly sought knowledge about more distant planets. Fortuitously, in the late 1970s, the largest planets in the solar system lined up on one side of the Sun, a phenomenon that occurs once every 176 years. A carefully timed spacecraft sent from Earth could take advantage of this configuration to make an unprecedented tour of the planets in a single voyage. Although the first nine spacecraft of the Pioneer series did not

---

[45] George Low statement, "Review of Recent Launch Failures," Hearings Before the Subcommittee on NASA Oversight, 15–17 June 1971, 8.

[46] Joseph Karth statement, ibid., 26.

[47] John E. Naugle statement, ibid., 61.

[48] Joseph Karth statement, ibid., 62.

Atlas-Centaur in flight just prior to Centaur second stage's separation from the Atlas first stage. (NASA Glenn Research Center unprocessed photo)

use the Centaur upper stage, Pioneers 10 and 11 needed the greater Centaur thrust capability to reach Jupiter and other points in the outer solar system.[49]

Pioneer 10 (AC-27) arrived at Pad 36A in December 1971. One month later, TRW shipped the Pioneer spacecraft to the Cape, and engineers began integrating the payload with the spacecraft. High upper altitude winds and a loss of facility electrical power caused the first delay, and continued unacceptable upper altitude winds caused two further postponements. Then, after last-minute tanking difficulties almost aborted the launch, the spacecraft lifted off on 2 March 1972, the beginning of one the longest voyages ever taken by a humanmade object.[50]

Pioneer 10 was unusual because, in addition to its second Centaur stage, it had a spin-stabilized Delta third stage with a solid rocket propellant motor (Thiokol TE-M-364-4), which

[49] Roger D. Launius, *NASA: A History of the U.S. Civil Space Program* (Florida: Krieger Publishing Company, 1994), 102. See also "AC-27 Launch Operations Pioneer F Mission," Box AC-25 to AC-32, Division Atlas/Centaur Project Office, NASA GRC Records.

[50] "AC-27 Flight Data Report," Launch Vehicles Division, Lewis Research Center, 1 June 1972, Box AC-25 to AC-32, Division Atlas/Centaur Project Office, NASA GRC Records.

allowed it to achieve an extremely high velocity of 32,000 miles per hour.[51] The Delta stage was mounted on a spin table at the top of Centaur. Thruster rockets spun this craft at 60 revolutions per minute, and explosive bolts separated it from Centaur after a 7-minute Centaur burn. To prevent Delta's reflected motor exhaust from damaging Pioneer, Centaur performed a retromaneuver to move it 25 feet from the Delta stage.[52] Delta ignition lasted for 44 seconds. Then another explosive bolt firing took place, with compressed springs pushing the Pioneer spacecraft away from the Delta stage. Throughout all of these complex maneuvers, the Atlas, Centaur, and Delta stages performed almost flawlessly. The Pioneer spacecraft entered into a successful Earth-escape orbit that required only a minor course correction. With Pioneer 10's successful launch bolstering the confidence of the launch vehicle team, they planned to replace the Atlas-Centaur computer system with one that was more advanced for Pioneer 11.

The Improved D-1A Centaur and Pioneer 11

NASA tested the improved Centaur D-1A with its new Teledyne avionics on Pioneer 11 (AC-30), one of the highest profile payloads of the United States. Although NASA had originally planned to make an Intelsat satellite the first test of the new avionics system, Comsat officials had objected. Everyone involved knew the risk—a failure would be a significant setback to the space program. Pioneer 11 had the same destination as Pioneer 10. It was again carried aloft by an Atlas, Centaur, and Delta vehicle combination. This redundancy was especially important because of the ambitious goals of the mission and the possibility that one of the three might fail.

Development of the D-1A improved Centaur began in 1968 with the goals of lowering costs associated with Centaur launches and increasing rocket reliability. There were twenty engineers assigned to the project, headed by Russ Dunbar. Long associated with the Centaur program, Vincent Johnson, now Deputy Associate Administrator for Space Science and Applications, stated to Congress that the reason for improving the Centaur was that many of its features were becoming outdated. He said, "It was recognized that the technology on which certain key systems of the Centaur was based was becoming relatively ancient."[53] Nearly $40 million was spent in creating the new Centaur, with $8 million in development costs for a new Teledyne computer.

The improved Centaur was tested by NASA in its new Spacecraft Propulsion Research Facility, or B-2, at the Plum Brook Station in Sandusky, Ohio. This unique facility consisted of a

---

[51] Pioneer 10 and AC-27 Press Kit, Box AC-25 to AC-32, Division Atlas/Centaur Project Office, NASA GRC Records.

[52] "Atlas/Centaur 27 Pioneer 10 Flash Flight report," Centaur Operations Branch KSC-ULO, 20 March 1972, Box AC-25 to AC-32, Division Atlas/Centaur Project Office, NASA GRC Records.

[53] Vincent L. Johnson statement, "Review of Recent Launch Failures," Hearings Before the Subcommittee on NASA Oversight, 15–17 June 1971, 83.

huge stainless steel vacuum chamber 38 feet in diameter and 55 feet tall–large enough to put an entire Centaur upper stage in the chamber, tank it with liquid hydrogen and liquid oxygen, and fire its engines. The B-2 facility could simulate the environmental extremes encountered at heights of approximately 100 to 125 miles and test Centaur's ability to restart its engines in the vacuum of space. The issues that particularly concerned engine restart included the necessary "chilldown" of the engine; making sure that liquid hydrogen, not hydrogen gas, entered the boost pumps; and ensuring that proper pressures in the engine propellant inlets were maintained. The liquid-nitrogen-cooled walls of the B-2 facility simulated the temperatures of 320°F found in space, while quartz lamp thermal simulators mimicked the intense heat of the Sun.

Checkout testing of the new facility began in October 1969 with the first successful hot firing of Centaur engines in the B-2 facility on 18 December 1970. Centaur was mounted vertically in the chamber with its nozzles facing down. After ignition, the engines fired through a water-cooled exhaust diffuser and into a deep spray chamber. Thousands of gallons of water were sprayed into the hot gas to cool it and turn it into steam. An exhaust system released the steam into the air while the heated water was returned to the spray chamber. Glen Hennings, chief of the rocket systems division at Plum Brook Station, recalled, "All systems performed properly in the test and we achieved most of our basic objectives."[54] Further testing of Centaur engines in the B-2 facility revealed that the RL10 engines could be run with a pressurized propellant feed system, thus eliminating the need for boost pumps.[55]

The most important element of the improved Centaur was its new Teledyne computer. James Patterson and Don Garman of NASA, along with David Geyer of General Dynamics, played key roles in the ultimate success of the new computer. In contrast to the serial drum-type Librascope computer it replaced, the new Teledyne computer was a true digital computer that could be reprogrammed and updated with relative ease. It had a 16,384-word memory (24-bit)–five times the memory of the earlier Librascope computer. The new software gave Centaur complete control of Atlas for the first time. The new computer had fifteen different software modules to manage functions such as navigation, guidance, autopilot, propellant utilization, attitude control, sequencing, telemetry, and data management for both vehicles.[56] Previously, separate mechanical or

[54] Glen Hennings, quoted in "First Test for New Facility: Centaur 'Hot-Fires' in B2," *Lewis News* (2 January 1970).

[55] See Steven V. Szabo, "Centaur Space Vehicle Pressurized Propellant Feed System Tests," NASA TN D-6876, October 1972.

[56] A. B. Yanke, "AC-30 Preliminary Flight Analysis Report GDCA-HAB73-019," Box AC-25 to AC-32, Division Atlas/Centaur Project Office, NASA GRC Records. Some of the more important modules included Navigation (NAV), which provided position, velocity, and acceleration data to the guidance system; Powered Guidance (PGUID), which provided data to optimize trajectory and determine engine cutoff time; Coast Guidance (CGUID), which determined the next main engine start; Powered Autopilot (PAUTO), which maintained Centaur's stability during main engine firings; Coast Phase Autopilot (CAUTO), which kept Centaur stable during the parking orbit; Propellant Utilization (PU), which maintained the desired ratio of liquid hydrogen and liquid oxygen in the tanks; and the Computer Controlled Vent & Pressurization System (CCVAPS), which used redundant tank-pressure sensors to maintain liquid-hydrogen and liquid-oxygen tank pressures. See "Titan IIIE/Centaur D-1T Systems Summary," September 1973, Joe Nieberding Personal Collection.

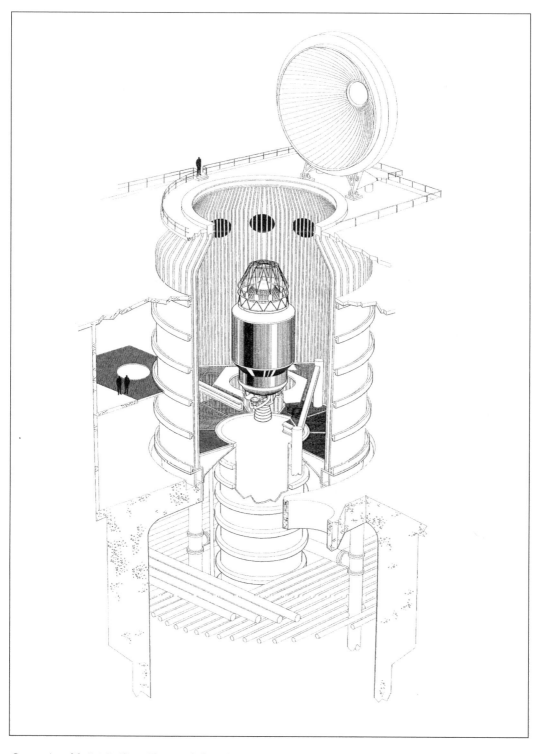

Cross section of the B-2 Facility at Plum Brook shows the Centaur rocket mounted for test. (CD-93-64366)

electronic units controlled functions such as propellant management and tank pressurization. Not only did the advanced computer system provide more integrated and customizable control over Atlas-Centaur, but it also allowed mission specialists greater flexibility in the design of a mission.[57] Instead of the redesigning hardware, the computer could be reprogrammed to prepare the Centaur for different tasks.[58]

The Teledyne guidance and navigation systems made possible one of the most significant contributions to Centaur D1-A capabilities. Because the computer could be programmed just before launch, a joint government-industry Centaur team created a new system to compensate for the winds the launch vehicle would encounter in its flight through the atmosphere. Winds are dangerous for launch vehicles because they create side forces that can either damage the vehicle or knock it off course. For the earliest Surveyor flights, Atlas-Centaur had used seasonal pitch and yaw programs based on historical data from sounding balloons. These early programs served as rough estimates of the winds that a vehicle might encounter on a typical launch day in summer, fall, winter, or spring. Frequently, weather balloons released on launch day revealed wind profiles quite different from the seasonal expectations. In 1966, the number of programs available expanded from four seasonal pitch and yaw programs to ten. Between 1966 and 1973, twenty-one Atlas-Centaur vehicles were launched using this protocol.

However, despite the power of the vehicle, it was still susceptible to upper atmosphere winds. If high jetstream winds on launch day did not somewhat match one of the ten predesigned pitch and yaw programs, then the entire launch had to be "scrubbed" or delayed until the winds cooperated. Delaying a mission was costly. What was needed was an approach that tailored the trajectory to actual real-time wind data on the launch day itself. Lewis engineers recalled that after Pioneer 10 was scrubbed twice in 1972 because of wind problems, Lewis Center Director Bruce Lundin demanded that the team come up with a solution by the time they launched Pioneer 11 thirteen months later. Engineers from the General Dynamics Aerodynamics Group, headed by Don Lesney, and a Lewis team, under Joe Nieberding, developed a revolutionary new real-time wind program called ADDJUST (Automatic Determination and Dissemination of Just Updated Steering Terms).[59] The ADDJUST system—an acronym coined by Frank Anthony—was incorporated into the D-1A Centaur launched in 1973.[60] It cost just $250,000 (roughly the same amount for one scrubbed launch).

The ADDJUST system dramatically increased the number of days available for launch. Prior to ADDJUST, about 43 percent of the launch dates between December and March were unavailable because of wind characteristics. After ADDJUST, even after allowing for uncertain-

[57] Interview with Roy Roberts by Virginia Dawson, 21 March 2001.

[58] Don Savage and Ann Hutchison, "Pioneer 11 to End Operations After Epic Career," NASA press release, 29 September 1995.

[59] Interview with Frank Anthony by Virginia Dawson, 6 June 2000.

[60] "Concept Review ADDJUST System to Design Booster Steering Programs During Preflight Launch Operations," 7 July 1972, Joe Nieberding Personal Collection.

Mariner Venus/Mercury 10 is prepared for encapsulation on 21 September 1973 at Kennedy Space Center. (NASA KSC_73P_0541)

ties in the actual winds encountered in flight, this percentage decreased to only 5 percent of launch days. The rest of the year, from April to November, Centaur achieved nearly 100-percent launch availability with respect to upper level winds. The ADDJUST system was used on all seven Titan-Centaur launches and has continued to play a vital role on all Atlas-Centaur launches since that time. After 1973, a launch with a Centaur upper stage would rarely be scrubbed due to an upper level wind problem.[61]

The process of implementing ADDJUST began 125 minutes before a scheduled liftoff. At this point, the Air Force released specially designed weather balloons into the Florida skies near the Eastern Test Range. As the balloons floated higher into the sky, they recorded wind velocity and direction at all altitudes, up to about 80,000 feet in about 60 minutes of rise time. After 20 minutes, these balloons began to transmit the data on the wind patterns to a ground-based computer. Five minutes later, using 300-baud modems, computers at General Dynamics in San Diego and Lewis Research Center in Cleveland received the data. Using two Fortran programs, their computers processed the data for 10 minutes. Then the General Dynamics computer transmitted pitch and yaw data to the Eastern Test Range in Florida, while the Lewis Research Center computer validated the approach. Ten minutes later, General Dynamics sent Lewis Research Center the current pitch and yaw data, and their computer in Cleveland validated the information for the next 10 minutes. After all the data were confirmed, they were sent to the onboard Centaur computer 15 minutes before the scheduled liftoff. To ensure that the data were accurate, Centaur transmitted the same data back to the computer at the Cape for verification.

All the Centaur engineers felt the anxiety of getting the improved Centaur ready on time. During tests of the new computer, its many glitches raised serious questions about reliability. Lewis engineer Ed Ziemba recalled that every time Teledyne delivered a computer, General Dynamics would test it and it would fail.[62] Up to 15 minutes before the launch of Pioneer 11, the computer was still giving the engineers trouble. Ground-based computers began spewing out strange data. Ziemba, who was assigned to the blockhouse, agonized over whether to postpone the launch. Because he was sure that the engineers had thoroughly checked out the new flight control system, he gave the go-ahead. His decision was vindicated by the perfect launch of Pioneer 11 on 5 April 1973. Atlas-Centaur (AC-30) cut through the Florida winds effortlessly with the ADDJUST system and placed its payload in its proper trajectory.

Pioneers 10 and 11 gathered unprecedented knowledge of the solar system. Pioneer 10 became the first spacecraft to travel through the asteroid belt, the first to make direct observations of Jupiter, and the first to venture outside the solar system. At a distance of 7 billion miles from Earth, it long held the distinction as the most remote human artifact. Pioneer 11, originally on a redundant path, adjusted its course to make observations of Jupiter as well as the first close observations of Saturn

[61] "ADDJUST Overview," Lewis Research Center, undated, Joe Nieberding Personal Collection.

[62] Interview with Ed Ziemba by Virginia Dawson, 19 May 1999.

in 1979. By 1995, the signal it sent back to Earth was so weak that scientists could only receive about 2 hours of data per month. In 1996, all communications fell silent and Pioneer 11 now drifts as a ghost ship headed for the Aquila constellation that it will reach in four million years.[63]

The First Two-Burn Planetary Mission: Mariner 10

Centaur's association with the Mariner program came to an end in 1973 with the Venus/Mercury launch (AC-34). Mariner 10's objective was a flyby of both Venus and Mercury–NASA's first two-planet mission. The complexity of this mission required that the new Centaur D1 use two burns in the launch of its interplanetary spacecraft. Although the two burns had been tested during the proof flights for Surveyor and used to launch Intelsat IV missions, Mariner 10 was the first planetary mission that required Centaur to restart its engines in space.

Deciding whether to use direct ascent or two burns depended on the geometry of Earth's location and the ultimate destination for the spacecraft. If Earth and the destination planet were aligned in a way that allowed the voyage to be made with just one propulsive push from the Centaur, direct ascent was selected. However, sometimes Earth and the destination planet were aligned in such a way that it was impossible to make a straight shot.

Prior to 1973, scientists had avoided the risk of the second burn and accepted the limitations of the direct ascent. The four previous Mariner launches to Venus and Mars were accomplished without the two-burn option. However, the alignment of the two innermost planets with Earth for Mariner 10 necessitated the more complicated mission profile. Although NASA could have used the more powerful Titan-Centaur for a direct ascent to Mercury, this option was more costly. Also, the greater speed of Titan-Centaur as it passed by the planets would have limited the time during which Mariner 10 could acquire data. Atlas-Centaur offered the best alternative. However, many scientists objected that using two burns might jeopardize the mission.[64]

Despite the complexity of the mission, when NASA launched Mariner 10 in November 1973, there were no unexpected problems or countdown holds. The Atlas phase ended 254 seconds into the launch. Centaur separated, and 10 seconds later, its engines fired. After 12 seconds, it jettisoned the nose fairing; 297 seconds later, the engines cut off and the rocket went into parking mode in a near-circular orbit of Earth. During its 1,540-second parking orbit, Centaur cruised into position, moving one-third of the way around Earth. The second burn began within 2 seconds of the predicted start. This sent Centaur, with its precious Mariner

---

[63] For information on the Pioneer 10 and 11 voyages, see William R. Corliss, *The Interplanetary Pioneers*, three volumes (Washington, DC: NASA SP-278 Government Printing Office, 1972); Richard O. Fimmel, William Swindel, and Eric Burgess, *Pioneer Odyssey* (Washington, DC: NASA SP-396 Government Printing Office, 1977); Elizabeth A. Muenger, *Searching the Horizon: A History of the Ames Research Center, 1940–1976* (Washington, DC: NASA SP-4304 Government Printing Office, 1985); Mark Washburn, *Distant Encounter: The Exploration of Jupiter and Saturn* (San Diego: Harcourt Brace Jovanovich, 1983); Tom Gehrels, ed., *Jupiter* (Tuscon: University of Arizona Press, 1976); Glenn S. Orton, "Pioneer 10," in *USA in Space*, 490–494; and Manfred N. Wirth, "Pioneer 11," in *USA in Space*, 495–499.

[64] Interview with William O'Neil by Virginia Dawson, 9 June 2000.

Atlas-Centaur 34 undergoing a tanking test on NASA Complex 36B at Cape Kennedy. This launch was the first dual-planet flight, Mariner Venus/Mercury 10. 19 September 1973. (NASA KSC_73P_0537)

payload, on a course opposite to the direction of Earth's orbital motion around the Sun. Velocity at cutoff was greater than statistical predictions because of a drifting Centaur gyro, but fortunately, Centaur's computer programming could compensate for this error and successfully place Mariner on its complicated trajectory. After 133 seconds–just as it entered an Earth-escape orbit–the Centaur main engines shut down for the final time. Centaur and Mariner stayed connected for another 95 seconds before they separated. After 515 seconds, Centaur performed a maneuver to get out of the way of the spacecraft.[65] In addition to requiring two burns, the Mariner trajectory was significant for another reason: it was the first time that a spacecraft used the gravitational attraction of one planet to catapult itself to another–a "feat requiring exceptional navigational accuracy and multiple trajectory corrections" to be performed by the trajectory analysis team.[66]

Mariner 10 reached Venus in February 1974. One month later, it made its first of three flybys of Mercury. After this first approach, Mariner 10 went into a solar orbit that permitted it to make two more encounters with the planet. It made its second Mercury encounter in September 1974, passing over the sunlit side of the planet and the southern polar region. Mariner 10 made its final pass by Mercury in March 1975–making its closest approach to the planet at 203 miles. It went silent eight days later when the fuel was depleted. Mariner 10 remains the only spacecraft to have visited Mercury. It made the first examination of the interplanetary space between Venus and Mercury and returned the first close observations of the cloud cover over Venus.[67]

A Journey to the Mysterious Planet: Pioneer Venus

While the previous two Pioneer spacecraft took long voyages away from the Sun and out into deep space, two additional Pioneer probes in the late 1970s were launched toward the second planet from the Sun. Venus, long known as the "mysterious planet," was most like Earth, with a similar mass, radius, and density. However, telescopic observations from Earth and data from the Mariner 10 probe suggested some striking differences. Temperatures on Venus approached 850 degrees, and its magnetic field was 1/3,000 of Earth's. Its rotational motion was 1/243 of Earth's and in the opposite direction. Finally, Venus was thought to be extremely arid, with a water

---

[65] "AC-34 Atlas-Centaur Flight Evaluation Report," Box AC-33 to AC-41, Division Atlas/Centaur Project Office, NASA GRC Records.

[66] Pioneer used gravity-assist as well, but although it was launched first, it was the second spacecraft to use the gravity-assist technique. See "Mariner Venus/Mercury 1973, Launch Operations and Flight Events," Box AC-33 to AC-41, Division Atlas/Centaur Project Office, NASA GRC Records.

[67] For further information on Mariner 10, see James A. Dunne and Eric Burgess, *The Voyage of Mariner 10: Mission to Venus and Mercury* (Washington, DC: NASA SP-424 Government Printing Office, 1978); Clark R. Chapman, *Planets of Rock and Ice: from Mercury to the Moons of Saturn* (New York: Charles Scribner's Sons, 1982); Bruce C. Murray and Eric Burgess, *Flight to Mercury* (New York: Columbia University Press, 1976); Robert G. Strom, *Mercury, the Elusive Planet* (Washington DC: Smithsonian Institution Press, 1987); and Donna Pivirotto, "Mariner 10," in *USA in Space*, 375–379.

content of 1/10,000 of Earth's biosphere. How could a planet so similar to Earth have surface conditions so radically different? Space scientists wanted to make a comparative study of Earth and Venus in the hopes of not only solving these tantalizing questions, but also shedding new light upon the evolution of our own planet.[68]

By the mid 1970s, the United States and the Soviet Union had sent a total of eight missions to study the planet. These included three Mariner flyby missions from the United States and five more complex missions by the Soviet Union with its Venera craft. The final two craft in this series (Venera 7 and 8 in 1971 and 1972, respectively) each landed on the surface of the planet and conducted measurements. Scientists in the United States believed that they were slipping behind their Soviet counterparts and sought to catch up. Although Pioneer Venus did not beat Soviet accomplishments, it served as the first United States mission to send both an orbiter and an atmospheric probe on separate launches.

NASA had originally chosen the Delta launch vehicle for Pioneer Venus. However, as scientists began designing their experiments, they found themselves severely constrained by Delta's weight capabilities. Either they could miniaturize and reduce the scope of their experiments, or they could spend an additional $10 million on an Atlas-Centaur launch vehicle. They soon realized that the funds saved by not having to miniaturize their experiments would offset Centaur's higher price tag.[69]

Atlas-Centaur (AC-50) launched the Pioneer Venus orbiter in May 1978.[70] Over the next fourteen years, the orbiter successfully carried out seventeen scientific experiments. It maintained an orbit around the planet until fuel depletion in August 1992 and disintegrated upon entering the planet's atmosphere. While Pioneer Venus yielded significant scientific results about the planet, these results were overshadowed by the excitement generated by a complicated Venus-bound spacecraft called the Multiprobe launched two and a half months later. Like the Pioneer Venus orbiter, the Multiprobe trajectory required a two-burn ascent with similar launch times and procedures. When the first orbiter arrived at Venus in December 1978, the Multiprobe was only five days behind.

The Multiprobe carried four atmospheric-entry probes to gather data on the Venusian clouds, atmospheric structure, and circulation pattern of the atmosphere. One of these probes managed to send back radio signals for an hour after impact with the surface of the

---

[68] "Atlas-Centaur D-1A AC-50 Launch Operations and Flight Events," Box AC-50 to AC-57, Division Atlas/Centaur Project Office, NASA GRC Records.

[69] Richard O. Fimmel, Lawrence Colin, and Eric Burgess, *Pioneer Venus* (Washington, DC: NASA SP-461 Government Printing Office, 1983), 26.

[70] O. Frank Spurlock, "Trajectory and Performance," as found in "Pioneer Venus Orbiter Mission Flight Data Report," Box AC-50 to AC-57, Division Atlas/Centaur Project Office, NASA GRC Records.

The High Energy Astronomy Observatory (HEAO-3) encapsulation in 1979. It was launched by Atlas-Centaur 53 on 20 September 1979. (NASA 8003540)

planet.[71] The Pioneer Venus Orbiter and Multiprobe missions attracted significant press coverage, as well as kudos from NASA Headquarters. In a letter to Larry Ross, NASA Administrator Robert A. Frosch recognized the "overwhelming success of Pioneer Venus activities" and contributions of the Lewis team. The program represented the "professionalism and dedication of many groups of people at NASA, in universities, and in industry."[72]

In Search of Cosmic Rays: The High-Energy Astronomy Observatory

The earlier OAO series of heavy science satellites were followed by a new version of the orbiting satellite observatory called the High-Energy Astronomy Observatory (HEAO), managed by Marshall Space Flight Center. When the HEAO program began in 1970, NASA planned to launch the new satellites with two Titan IIIC boosters. In 1973, TRW won a contract to build the spacecraft, but budget cuts threatened cancellation of the program. Marshall was forced to scale back the program and reduced the weight of the satellites from 18,000 to 6,000 pounds so that they could be carried aloft by Atlas-Centaur. At that time, the satellites represented the heaviest mission attempted by an Atlas-Centaur.

The objective of the new class of HEAO spacecraft was to locate, identify, and analyze celestial high-energy radiation sources.[73] HEAO spacecraft were designed to survey the sky for x rays, investigate the shape and structure of these x-ray sources, measure gamma-ray flux and determine source locations, and increase the understanding of various cosmic phenomena.[74] With a series of three launches planned (all aboard the improved Centaur D1 with a single burn), scientists hoped that the HEAO would help them begin to uncover answers to "the most intriguing mysteries of the universe—pulsars, black holes, neutron stars, and supernovae."[75]

The HEAO program represented the first use of new in-flight Centaur retargeting software. This feature could be used if Centaur lost power in flight. The software automatically changed the aiming point for the vehicle to a lower orbit. Although this compromised the flight, it prevented complete mission failure. No HEAO required the retargeting software because all three Centaurs performed as expected; nevertheless, the advance was an important software feature for future missions.

---

[71] For further information on the Pioneer Venus missions, see Eric Burgess, *Venus: An Errant Twin* (New York: Columbia University Press, 1985); D. M. Hunten et al., eds., *Venus* (Tuscon: University of Arizona Press, 1983); and Robert J. Paradowski, "Pioneer Venus 1" and "Pioneer Venus 2," in *USA in Space*, 500–508.

[72] Robert Frosch to Larry Ross, 10 April 1979, Box AC-50 to AC-57, Division Atlas/Centaur Project Office, NASA GRC Records.

[73] J. R. Brown, "RL10A-33 Engine Performance Summary for Atlas Centaur Flight AC-45," Box AC-42 to AC-49, Division Atlas/Centaur Project Office, NASA GRC Records.

[74] Edwin Muckley, "HEAO Mission Objectives," from the Director of Launch Vehicles's report to the Director, Box AC-42 to AC-49, Division Atlas/Centaur Project Office, NASA GRC Records.

[75] "Atlas-Centaur D-1A AC-52," Box AC-50 to AC-57, Division Atlas/Centaur Project Office, NASA GRC Records.

The mission of the first HEAO, launched on an Atlas-Centaur (AC-45) on 12 August 1977, was to map the sky broadly for x and gamma rays. The countdown was interrupted for electrical storms and the brief failure of a payload ground computer. Then shrimp boats encroached upon the range safety area in the Atlantic Ocean. Once the computer was fixed, the storms over, and the shrimp boaters banished, liftoff occurred with 13 minutes to spare. Centaur completed its single-burn ascent and placed the 5,581-pound HEAO-1 in its designated circular orbit 276 miles above Earth.[76] After separation from the HEAO, Centaur began a "deflection turn" 11 seconds later to minimize any possible chance of contaminating the HEAO. This turn was a 90-degree pitch and yaw maneuver in which Centaur pointed its engines towards the Sun. At the same time, Centaur engaged in a 45-degree roll. When Centaur finished repositioning, the propellant tank's settling and venting sequence was initiated. This procedure further increased the distance between Centaur and the HEAO. The final Centaur "blowdown" phase began 74 minutes after launch and continued for 305 seconds, lowering it below the HEAO orbit.[77] (During blowdown, any residual propellants are vented from the Centaur tanks before the vehicle reenters the atmosphere.) HEAO-1 conducted nearly three full celestial surveys and discovered 1,500 new sources of x rays in the sky.

NASA launched the second spacecraft in this series, HEAO-2, in November 1978. The Atlas-Centaur (AC-52) launch sequence was similar to that of HEAO-1, but after payload separation, Centaur carried out two experiments designed to improve its own future performance. The first experiment occurred 106 minutes into the flight. During the first orbital pass over Florida, Centaur began a special spin-up experiment to demonstrate its ability to spin a commercial spacecraft to 12 degrees per second before payload separation.

The second experiment occurred 109 minutes into the flight. Using newly implemented Centaur tumble-recovery software made possible by the new Teledyne computer, the vehicle simulated a severe tumble of 329 degrees of pitch and 584 degrees of roll. The new software allowed a controlled steering maneuver to realign Centaur. This was the first NASA vehicle to have tumble-recovery software encoded into its onboard computer software.[78]

The operational objectives of HEAO-2 were similar to those of the first craft in the series. Again it searched the universe for x, gamma, and other cosmic rays and mapped their locations. However, the HEAO-2 was larger and contained much more sophisticated observational equipment than its predecessor. At 6,866 pounds, it became the heaviest object ever lifted into space by an Atlas-Centaur, and it orbited Earth at 333 miles. The centerpiece of all the

---

[76] This was the first mission in which Centaur delayed its usual separation time because engineers wanted telemetry coverage over Ascension Island. "AC-45 Preliminary Analysis report," 1 September 1977, Box AC-42 to AC-49, Division Atlas/Centaur Project Office, NASA GRC Records.

[77] "Atlas Centaur Flight Evaluation AC-45," Box AC-42 to AC-49, Division Atlas/Centaur Project Office, NASA GRC Records.

[78] "Atlas-Centaur Postflight Analysis AC-52," Box AC-50 to AC-57, Division Atlas/Centaur Project Office, NASA GRC Records.

instrumentation was a large x-ray telescope that had the ability to isolate and point directly at the most interesting stellar emanations that HEAO-1 identified in the sky.[79] HEAO-2 made over five thousand specific observations and also discovered that both Earth and Jupiter emit x rays. HEAO-3 contributed to a better understanding of interstellar matter by detecting cosmic-ray particles and gamma-ray photons.

The final launch in the HEAO series occurred on 20 September 1979 with Atlas-Centaur (AC-53). At 6,321 pounds, HEAO-3 was actually significantly lighter than HEAO-2. Its objectives were similar to those of HEAO-1.[80] It was the last mission of Atlas-Centaur to launch a low-Earth-orbiting satellite until the commercial era of the 1980s.

Despite all of their mission complexity, almost all of the Atlas-Centaur missions succeeded during the busy decade of the 1970s. The team lost one Orbiting Astronomical Observatory and one Mariner mission, caused by the malfunction of a tiny integrated circuit in the Centaur autopilot. Launches of two of nineteen Intelsat commercial satellites also failed. These setbacks did nothing to lesson the resolve or growing confidence among the government-industry Centaur team. They continued a tradition of innovation on Centaur with the upgrade to its computer system. Centaur reliability was greatly enhanced by the ability to ground-test its restart capability in a new vacuum facility completed at Lewis Research Center's Plum Brook Station in the early 1970s.

The early successes of the Atlas-Centaur planetary program paved the way for even more ambitious missions in the service of heavy science. The Titan-Centaur program, with voyages to the Sun, Mars, and the outer planets, continued the legacy of the Atlas-Centaur program. Titan III-Centaur sent spacecraft to probe deeper into the secrets of Earth's planetary neighborhood and enabled scientists to learn more about the evolution of Earth itself. Questions about the future of expendable rockets seemed not to dampen the enthusiasm of the engineers and scientists involved in the planetary science program for developing an even more powerful launch vehicle—the giant Titan-Centaur.

---

[79] "Atlas/Centaur D-1A AC-52 HEAO-B Flight Data Report," February 1979, Box AC-50 to AC-57, Division Atlas/Centaur Project Office, NASA GRC Records.

[80] "Atlas-Centaur D-1A AC-53 HEAO-C Mission," Box AC-50 to AC-57, Division Atlas/Centaur Project Office, NASA GRC Records. For further information about the HEAO satellites, see James Cornell, Wendell Johnson, and Carroll Dailey, eds., *High Energy Astronomy Observatory* (Washington, DC: NASA EP-167 Government Printing Office, 1980); Wallace H. Tucker, *The Star Splitters: The High Energy Astronomy Observatory* (Washington, DC: NASA SP-466 Government Printing Office, 1984); Dave Dooling, "Window on Violent, Cataclysmic Universe Closes," *Space World* (May 1982): 16–17; and Dooling, "The High-Energy Astronomical Observatory," in *USA in Space*, 253–256.

Chapter 5

The Giant Titan-Centaur

"But when Viking came along—it was the first Mars lander—it was too heavy for the Atlas-Centaur so we had to go with a bigger booster, Titan . . . . All that effort spent to integrate the Centaur to the Titan, [and] we only launched it seven times, because by the time we launched Voyager the Shuttle was on the horizon and it was going to displace expendable launch vehicles."

—Joe Nieberding, Lewis Research Center

While Atlas-Centaur was proving its heavy lift capability, a giant launch vehicle that delivered even more power emerged in 1974–Centaur coupled with a Titan IIIE booster. Titan-Centaur was capable of lifting unprecedented heavy science missions. The heaviest Atlas-Centaur planetary launch was the 2,201-pound Mariner 9 spacecraft; Titan-Centaur would more than triple this capability with the launch of the 7,767-pound Viking probe. Wernher von Braun, once skeptical that a liquid-hydrogen upper stage was even feasible, showered the new Titan-Centaur with high praise shortly before its first launch. In *Popular Science*, he announced the flightworthiness of a "new generation of heavyweight interplanetary spacecraft."[1] Standing 160 feet tall, the Titan-Centaur launch vehicle would give scientists the capability to learn more secrets about the solar system than ever before. With a hint of regret, von Braun further wrote that the Titan "has the prospect of becoming NASA's largest launch vehicle in active use since our two remaining Saturn V's are unassigned and mothballed." Wernher von Braun's praise was justified. Bigger and more powerful than Atlas-Centaur, Titan III-Centaur proved to be one of the most powerful and flexible launch vehicle systems ever designed.

Titan-Centaur launched key science missions to the Sun, Mars, and Jupiter with the Helios, Viking, and Voyager probes. Ironically, despite its dramatic impact on the planetary space program, NASA gave this Titan III-Centaur a life of only seven launches. Even before the Titan-

---

[1] Wernher von Braun, "Our Biggest Interplanetary Rocket," *Popular Science* (July 1974): 62.

The Seven Launches of Titan-Centaur for the Space Program (1974–1977)				
Launch Date	Mission	Vehicle	Payload	Objective
11 February 1974	Proof Flight	TC-1	7,985 lb	Simulated Viking launch. Sphinx spacecraft. Failed.
10 December 1974	Helios 1	TC-2	1,093 lb	Sent Helios 1 to the Sun.
9 September 1975	Viking 2	TC-3	7,767 lb	Sent Viking 2 to Mars.
20 August 1975	Viking 1	TC-4	7,764 lb	Sent Viking 1 to Mars.
15 January 1976	Helios 2	TC-5	1,102 lb	Sent Helios 2 to the Sun.
20 August 1977	Voyager 2	TC-7	2,168 lb	Sent Voyager 2 to Jupiter.
5 September 1977	Voyager 1	TC-6	2,171 lb	Sent Voyager 1 to Jupiter.

Centaur had a chance to prove itself, its grave was already being prepared. As the new launch vehicle rolled out to the launch pad for its inaugural flight, the expendable launch vehicle was marked as a dying breed. John Noble Wilford wrote in the *New York Times*, "The mammoth rocket, the Titan III-Centaur, is expected to be the last new launching vehicle to be developed by the National Aeronautics and Space Administration until the advent of the reusable space shuttle."[2] Despite its uncertain future, Centaur, mated so elegantly and powerfully with Titan, would make possible the acquisition of significant new knowledge of the solar system.

Integrating Titan and Centaur

In the early 1960s, no one had predicted that Titan and Centaur would mate. NASA's long-range launch vehicle plan was to continue to use Atlas-Centaur until a reusable launch system or a nuclear-powered upper stage could be developed. The escalation of the Vietnam conflict and President Johnson's War on Poverty disrupted these plans. Congress drastically reduced expenditures for the civilian space program, and the development of a reusable launch vehicle was put on hold. Because NASA needed a launch vehicle more powerful than Atlas-Centaur to send heavier unpiloted planetary probes like Viking and Voyager into space in the 1970s, NASA began in 1967 to consider the possibilities of mating a Centaur upper stage with the giant Titan III.[3]

[2] John Noble Wilford, "Rocket for Exploration of Planets Rolled to Pad for 'Proof Flight,'" *New York Times* (3 October 1973).

[3] Linda Neuman Ezell, *NASA Historical Data Book, Volume III* (Washington, DC: NASA, SP-4012, 1988), 38.

Titan, an Air Force launch vehicle developed in the mid-1950s as an ICBM by the Glenn L. Martin Company (later Martin Marietta Corporation), was a behemoth. But the tons of liquid-oxygen oxidizer and RP-1 kerosene it required meant that it could not be readied for flight rapidly enough in the event of a military emergency. To solve the propellant problem, the Martin Company developed a new two-stage version called the Titan II that used storable, hypergolic propellants. These propellants did not require the elaborate tanking procedures of Titan I. Titan II's power promised an unprecedented level of performance and cost-effectiveness as a launch vehicle for NASA. Titan II's first space mission to lift a two-person Gemini capsule was considered a triumph.[4]

The Titan III program began in 1962 when the Department of Defense authorized the development of a launch vehicle for heavy military satellites. Titan III had two solid rocket strap-ons for additional thrust. The Air Force tested the ability of Titan IIIA and IIIC to lift 13,000-pound payloads into low-Earth orbit in 1964. One of the most significant additions to Titan was an upper stage rocket called Transtage intended to carry multiple satellites into different orbits on a single mission. Because Transtage was plagued with development problems, NASA chose Titan-Centaur as an upper stage for its heavy science missions.

Lewis engineers confronted the daunting prospect of integrating Centaur with Titan III. Andrew Stofan, former head of the Propellant Systems Section, managed the Titan-Centaur project office at Lewis. His counterpart at General Dynamics, Russ Thomas, led the development of the Centaur D-1T and managed the industry team supporting the seven Titan III-Centaur D-1T planetary launches. John Neilon, who headed Unmanned Launch Operations at Kennedy Space Center between 1970 and 1975, recalled, "It was a distinct break from our way of doing business on Atlas-Centaur and Delta."[5] All three organizations had to learn how to deal with a new contractor, Martin Marietta, as well as the Air Force, which owned both Titan and the launch pad.

Titan was launched from the Integrate, Transfer and Launch (ITL), its own launch facility. The ITL consisted of the Vertical Integration Building (VIB), the Solid Motor Assembly Building (SMAB), and two launch pads, 40 and 41. Pad 41 was modified for Titan-Centaur. All launches for the Titan-Centaur were from Launch Complex 41, owned and controlled by the Air Force.

Prior to reaching the launch pad, the vehicles went through several check stages. The Titan was assembled and tested at Martin Marietta's Denver, Colorado, plant. The Titan core vehicle was then sent by plane to Florida. United Technologies built the solid rocket motors in

---

[4] Frank Winter, *Rockets Into Space* (Cambridge: Harvard University Press, 1990), 91–93. See also Robert L. Perry, "The Atlas, Thor, Titan, and Minuteman," in *The History of Rocket Technology*, ed. Eugene M. Emme (Detroit: Wayne State University Press, 1964), 160; and Barton C. Hacker and James M. Grimwood, *On the Shoulders of Titans: A History of Project Gemini* (Washington, DC: NASA SP-4203, 1977).

[5] Communication from John Neilon to the authors, 7 July 2002.

Sunnyvale, California, and shipped them to Florida by rail. The Aerospace Division of General Dynamics assembled the Centaur in San Diego, California, with final preparations taking place at the Kearny Mesa facility. From there, Centaur was taken to Miramar Naval Air Station, with a final flight on a C-5A plane to Kennedy Air Force Station. At Kennedy, Titan was moved to the VIB, where it was enclosed in a service structure. After checkout, a railroad locomotive pulled the core vehicle, consisting of the two Titan stages, on tracks to the SMAB, where the two solid rocket motors were attached to the Titan core vehicle. The Titan-Centaur was moved by rail one last time to Complex 41 to be mated with its payload.[6]

Since Pad 41 did not have a blockhouse, checkout and launch functions were carried out from the Launch Control Center in the VIB, located several miles from the launch site. Neilon wrote, "Distance, rather than concrete provided safety. The same distance was also something of an inconvenience; one did not simply walk out to the pad as one could on Complex 36, but we got used to it. In summary, being different didn't mean wrong or inferior, just different. We all worked together and made things work."[7]

When erected on the launch pad, Titan-Centaur was an impressive sight. Together, the two vehicles stood 160 feet tall and were 10 feet in diameter, with a 14-foot-diameter shroud. Flanking the Titan on either side were the two 10-foot Solid Rocket Motors (SRMs) built by the Chemical Systems Division of United Technologies. The SRMs were capable of 2.4 million pounds of thrust. Each weighed 500,000 pounds and stood 85 feet tall. When the SRMs (considered stage 0) burned out, they were jettisoned and the Titan's first and then its second stage took over. These were called the "core vehicle." Both burned liquid Aerozene-50 fuel and red fuming nitric acid (RFNA) as an oxidizer. The first stage, 73 feet long and 10 feet in diameter, provided 470,000 pounds of thrust. The 23-foot second stage delivered 100,000 pounds of thrust. Aerojet manufactured both core vehicle engines.

The Centaur D-1T (stage 3) was 23 feet in length and weighed 39,000 pounds, including its liquid-oxygen and -hydrogen propellants. Its two Pratt & Whitney main engines each provided 15,000 pounds of thrust.[8] Many features of the Centaur D-1T for Titan were the same as those of the Centaur D-lA for Atlas. For example, Centaur remained the "brains" because of its Teledyne Digital Computer Unit with its 16-K, twenty-four-bit capacity. The computer had a twenty-five-term instruction set, plus additional input and output instructions. The Titan version offered greater computational reserve memory and input/output functions.[9]

---

[6] "Titan IIIE/Centaur D-1T Systems Summary," September 1973, Joe Nieberding Personal Collection.

[7] Ibid.

[8] John F. Kennedy Space Center, "Titan/Centaur," press release, 20 August 1977.

[9] "Titan IIIE/Centaur D-1T Systems Summary," September 1973, Joe Nieberding Personal Collection. Unlike Atlas, which was a one-and-a-half-stage booster, Titan was a two-stage rocket (stage 1 and stage 2) with two solid rocket motor strap-ons, called SRMs (stage 0). Centaur became the third stage and the Delta TE-364 the fourth stage.

The D-1T Centaur had an updated electrical system and a totally redesigned shroud assembly called the Centaur Standard Shroud (CSS), manufactured by Lockheed Missiles and Space Company.[10] It required extensive testing in the Space Power Facility at NASA's Plum Brook Station. The shroud enclosed both Centaur and the payload, serving as environmental protection for the payload on the ground and during the first few minutes after launch. It consisted of two halves wrapped around Centaur and held in place by eight compressed springs mounted in pairs. The shroud had a venting system that minimized the differences in structural pressures between the vehicle and spacecraft during flight through the atmosphere.[11] Once the vehicle left the atmosphere, the springs forced the halves to separate and the shroud was jettisoned.[12]

The Titan-Centaur used a thermal radiation shield to reduce the heat that Centaur's liquid-hydrogen fuel absorbed while it coasted in space. The radiation shield enabled Titan-Centaur to coast in space for longer periods than Atlas-Centaur, which had a maximum coast period of about 30 minutes. Titan-Centaur could coast for over 5 hours, providing improved synchronous-orbit capability and extended launch windows. Although Titan-Centaur payloads never needed these extended coast periods, these capabilities were tested on the two Helios missions. Technical information learned from these important experiments was later used in the development of Titan IV-Centaur and the commercial version of Centaur in the 1990s.

Integrating Centaur with Titan proved to be a significant technical challenge. When Seymour Himmel, then Associate Director for Rockets and Vehicles at Lewis, took charge of the integration of the two vehicles, he found that the procedure was flawed. There was a lack of clear and effective technical communication between the two major aerospace companies involved. He directed that a common database and set of definitions be established, documented, and maintained as the governing Interface Control Drawings (ICD) for the integration activity. This was no minor task because the two companies had differing cultures and design processes.[13]

In addition, he directed that complete structural and structural-dynamics analyses be conducted for the combined vehicle. This effort underscored the nature of the integration problems because the two companies did not use a common set of definitions for such analyses. Once a "dictionary" was developed and documented, the process went forward much more smoothly.[14]

---

[10] S. V. Szabo, Jr., and L. J. Ross, "Titan/Centaur 1 Post-Flight Evaluation Report," April 1975, Box TC-1 TC-2 TC-3 Records, Division Atlas/Centaur Project Office, NASA GRC Records.

[11] K. A. Adams, "Helios B Flight Data Report," Box TC-5, TC-6 Records, Division Atlas/Centaur Project Office, NASA GRC Records.

[12] Centaur Operations Division, Kennedy Space Center, "Titan-Centaur-2 Helios 1 Field Report," 6, Box TC-1 TC-2 TC-3 Records, Division Atlas/Centaur Project Office, NASA GRC Records.

[13] Interview with Seymour Himmel by Virginia Dawson, 1 March 2000, and his letter to her, dated 14 December 2002.

[14] Interview with Seymour Himmel by Virginia Dawson, 1 March 2000.

An early problem was the difference between the diameters of the Titan core and the Centaur shroud; the latter was four feet wider. The solution was to taper the Centaur shroud at the junction point between the two rockets. This created a bulge, which gave the launch vehicle its unique hammerhead shape. A more difficult problem resulted from the different temperatures of the two rockets. Heat transfer from Titan's extremely hot engines might compromise the very cold temperatures needed to keep Centaur's cryogenic propellants from becoming gaseous during launch. John Gossett, Centaur Operations division chief at the Cape, told Craig Covault of *Aviation Week* that to design a system with just the right amount of insulation without sacrificing weight required a major design effort. Installing insulation in the shroud and bulkhead between the stages satisfied the different thermal requirements.[15]

In addition to the technical challenges, dealing with the different cultures of the Atlas and Titan programs proved a challenge for Lewis project managers. Air Force personnel were rotated after a few years, a practice that prevented them from forming close personal relationships with their counterparts at Martin Marietta. In contrast, years of working together had produced a seamless relationship between the Lewis and General Dynamics people.[16]

Karl Kachigan, chief engineer of General Dynamics, recalled that the main problem was Air Force procedures that accounted for Martin Marietta's glacial slowness in responding to requests.[17] In addition to contending with the problem of protecting the proprietary technology of two rival aerospace companies, Lewis engineers also found it necessary to persuade Martin Marietta to use one of the innovations pioneered by General Dynamics–the ADDJUST wind program, developed shortly before the first Titan-Centaur launch. Once installed, the new wind technique worked flawlessly for every Titan-Centaur launch.

Before a launch, in addition to the prescribed launch checkout procedures, another set of unofficial preparations typically occurred. These were various superstitions, traditions, and rituals that NASA engineers brought with them to the Cape. For example, going back to Surveyor days, the launch vehicle team from Lewis Research Center all had to eat at Ramon's at Cocoa Beach, and everyone had to order the Caesar salad. Joe Nieberding recalled, "I never liked Caesar salad, but who was I to say, being new to the program at the time. We had a great success record, and we just didn't want to change anything."[18] Since Titan was an Air Force rocket, Air Force rituals now entered the mix. Key personnel from the mission would gather on the beach the night before a launch. A quiet, somber, reverent mood more befitting a séance room than an engineering/military gathering would settle over the group. A designated individual took a

---

[15] Craig Covault, "Titan 3E to Fill Gap in Space Boosters," *Aviation Week and Space Technology* (17 September 1973): 96; and Craig Covault, "First Titan 3," *Aviation Week and Space Technology* (4 February 1974): 52–55.

[16] Interview with Mike Benik by Virginia Dawson, 9 November 1999.

[17] Interview with Karl Kachigan by Virginia Dawson, 7 June 2000.

[18] Interview with Joe Nieberding by Virginia Dawson and Mark Bowles, 15 April 1999.

long knife and buried it in the sand with the knife pointing in the launch direction. "These were things you had to do," Nieberding recalled. "It's hard to believe, when you look back."[19] Other engineers had more private superstitions and rituals they performed before each launch. General Dynamics chief engineer Karl Kachigan admitted, "Whenever the launch was at T-minus 18 seconds, I would always cross my fingers. One of the fellows took a picture of that once and he gave it to me. He said, 'We don't have a chief engineer, we've got a witch doctor!'"[20]

An Inauspicious Beginning: The Proof Flight

Before the first missions to the Sun and Mars could be flown, a Proof Flight, designated as Titan-Centaur 1 (TC-1), was necessary to demonstrate the flightworthiness of the vehicle. The Proof Flight would have the same trajectory as the Viking mission to Mars that was scheduled to be launched in 1975. The Proof Flight was to carry the Viking Dynamic Simulator (VDS), a model of the Viking spacecraft which was not intended to actually separate from Centaur. Its role was simply to determine flight loads to which the real Viking would be subjected during its launch.

Although the Proof Flight was not originally planned to carry any payload, Lewis engineers responsible for the launch persuaded their colleagues at the Center to send an experimental package on it. Known as the Space Plasma High Voltage Interaction Experiment, or SPHINX, it was designed to measure how space plasmas interacted with high-voltage surfaces on the spacecraft. Lewis engineers planned to use this information to design better high-voltage systems to operate in space environments. The year-long SPHINX mission would gather sample data from many different types of plasma particles.[21]

The Proof Flight was a disaster.[22] After a normal launch countdown on 11 February 1974 (except for a 45-minute hold resulting from questionable data from the booster's hydraulic systems), the first Titan-Centaur lifted off at 9:48 a.m. Titan jettisoned its shroud and Centaur separated, but at this point, the Centaur engines failed to start. Without power, Centaur went into a freefall and was destroyed by Range Safety 748 seconds after liftoff.[23] While the loss was devastating for the members of the launch team, it was even more tragic for those involved in the Lewis SPHINX program. "They were looking for a ride," Joe Nieberding recalled. "And we said, 'We have this perfectly good rocket. It will be a whole Lewis thing with a Lewis payload

---

[19] Ibid.

[20] Virginia Dawson interview with Karl Kachigan, 7 June 2000.

[21] "Proof Flight, Launch Operations and Flight Events," Box TC-1 TC-2 TC-3 Records, Division Atlas/Centaur Project Office, NASA GRC Records.

[22] S. V. Szabo, Jr., and L. J. Ross, "Titan/Centaur 1 Post-Flight Evaluation Report," April 1975, Box TC-1 TC-2 TC-3 Records, Division Atlas/Centaur Project Office, NASA GRC Records.

[23] Kennedy Space Center report, "Titan/Centaur-1 Proof Flight (SPHINX) Field Report," Box TC-1 TC-2 TC-3 Records, Division Atlas/Centaur Project Office, NASA GRC Records.

and a Lewis rocket.' They said, 'Great!' We put them in the water and they never forgave us."[24] Despite the failure, many of the evaluation criteria for the Proof Flight were met.[25] The Proof Flight demonstrated the structural integrity of the new Centaur and its capability to jettison the Centaur Standard Shroud.[26]

The reasons for the failure of the Centaur engines were not immediately clear even after a detailed investigation revealed that the problem had occurred in the boost pumps.[27] The Failure Review Board, chaired by Seymour Himmel, could not pinpoint the exact cause of the failure of the Centaur boost pumps. The Board concluded that the two most probable causes were either 1) freezing of the hydrogen-peroxide-generated steam as it expanded through the boost-pump turbine or 2) ingestion of a foreign object that jammed the pump, preventing it from rotating. No obvious source of such an object could be identified, since no change had been made in the way the Centaur propellant tanks were built. Tests conducted at the B-2 facility at the Plum Brook Station proved inconclusive. The best corrective action that could be devised was a verification procedure to make sure that the boost pumps were free to rotate just before liftoff. Subsequent launches were successful despite the fact that the root cause of the Proof Flight failure had not been indisputably determined.

It took four years to solve the mystery. At General Dynamics, an employee that everyone called the "little old winemaker" had been, for a long time, the only one to install the "clip" used to anchor the propellant-utilization probe to the wall of the Centaur oxygen tanks. When he retired, the new person assigned the task had great difficulty in installing the clip using the materials and procedures specified in the shop instructions. Frequently, the clip fell off, but after several trials, he apparently succeeded in securing it. This difficulty was reported to the "little old winemaker," who confirmed the difficulty of anchoring the clip to the wall of the tank. He had concluded that the length of the rivet specified in the drawings was too short, so he had always obtained and used a longer rivet that worked much better. Unfortunately, he had never reported the problem or his solution. Since the new mechanic had installed the probe for the proof flight, it was concluded that the clip had fallen off and jammed the boost pump.[28] What was ironic was that none of the new Titan-Centaur components were at fault and the failure was part of the old, proven Centaur. Even though the first Titan-Centaur failed, both industry and government

[24] Interview with Joe Nieberding by Virginia Dawson and Mark Bowles, 15 April 1999.

[25] Craig Covault, "Centaur Failure Stirs Minimal Concern," *Aviation Week and Space Technology* (18 February 1974): 20.

[26] A. B. Yanke, "TC-1 Flight Data Preliminary Analysis," Box TC-1 TC-2 TC-3 Records, Division Atlas/Centaur Project Office, NASA GRC Records.

[27] General Dynamics report, "Titan/Centaur Flight Evaluation TC-1," Box TC-1 TC-2 TC-3 Records, Division Atlas/Centaur Project Office, NASA GRC Records.

[28] Interview with Seymour Himmel by Virginia Dawson, 1 March 2000.

engineers learned an enormous amount about the new rocket system. The next six Titan-Centaur launches were flawless. NASA's most ambitious probes to the Sun, Mars, Jupiter, Saturn, Uranus, and Neptune were hailed as important contributions to space science.

A German Partnership To Explore the Sun

In 1969, the United States and the Federal Republic of Germany (West Germany) agreed on their first joint space project to send two probes to explore the mysteries of the Sun, the star that has dominated human interest in the skies for millennia. These probes would provide unprecedented observations of the Sun, reveal important data about the relationship between the Sun and Earth, and also gather key data about the solar wind, magnetic and electric fields, cosmic rays, and cosmic dust.[29] The spacecraft would also test Einstein's theory of general relativity.[30] The mission was called Helios, named in honor of the ancient mythological Greek god of the Sun.

Originally, the Germans had planned to use Atlas-Centaur, but as the design evolved, glass plates and mirrors had to be added to dissipate the intense heat generated by the Sun. These additions brought the craft's weight close to the limits of what the Atlas-Centaur could lift. The Americans convinced the Germans that by using the untested Titan-Centaur, they would not have to compromise the scientific objectives of the mission by eliminating valuable scientific equipment. The Americans also promised the Germans that the second Helios flight would occur after the Viking Proof Flight. Reassured that any system flaws would be worked out by the time Helios was launched, the Germans agreed to use the untried vehicle. They were willing to tolerate these risks because of the payload weight opportunities that Titan-Centaur offered.[31]

Lewis Research Center managed the launch vehicle, Goddard Space Flight Center directed the overall mission, and Kennedy Space Center managed operations at the launch site. In Germany, the Federal Ministry for Research and Technology (Bundesminister für Forschung und Technologie) provided the spacecraft. The Gesellschaft für Weltraumforschung controlled all the technical facets of their construction. It also ensured that the prime contractor for the spacecraft, Messerschmidt-Boelkow-Blohm, as well as the other German contractors, worked together to deliver the probes to NASA on schedule. Each Helios was a relatively short, sixteen-sided, cylindrical spacecraft with two solar arrays attached to its front and aft ends. Each Helios carried seven German scientific experiments. The United States provided three additional experiments on each probe, along with tracking data from NASA's Deep Space

---

[29] See Karl Hufbauer, *Exploring the Sun: Solar Science Since Galileo* (Baltimore and London: The Johns Hopkins University Press, 1991); and "Titan-Centaur D-1T, TC-2, Launch Operations and Flight Events," Box TC-1 TC-2 TC-3 Records, Division Atlas/Centaur Project Office, NASA GRC Records.

[30] Glenn Research Center report, "Titan-Centaur D-1T TC-5 Launch Operations and Flight Events," Box TC-5, TC-6 Records, Division Atlas/Centaur Project Office, NASA GRC Records.

[31] Robert S. Kraemer, *Beyond the Moon: A Golden Age of Planetary Exploration 1971–1978* (Washington: Smithsonian Institution Press, 2000), 87.

A Helios spacecraft prototype is encapsulated in its payload fairing at Kennedy Space Center on 27 September 1974. This work was in preparation for a future Titan-Centaur flight. (NASA KSC_74P_0222)

Network (DSN). Most importantly, the Americans supplied the launch vehicles for these missions—two Titan IIIE/Centaur/Delta TE-364 rockets.

As the launch approached, however, tension grew in the relationship between the Germans and the Americans. After the Viking Proof Flight failed, the Germans feared that NASA was simply using their Helios mission to test the Viking launch profile before launching the expensive Viking spacecraft.[32] Specifically, they objected to using their mission to test a two-burn capability needed for Viking when Helios did not require it. If the restart were not achieved, this high-profile international mission would be stranded in low-Earth orbit.

Other concerns focused on the Centaur Teledyne digital computer. A circuitry problem within the computer was discovered in July 1974. During qualification testing, engineers found that 1 in every 10,000 microelectronic modular assemblies failed. Since Centaur contained 2,400 modules, this became a very serious concern. There was a possibility that the Helios mission, as well as a scheduled Atlas-Centaur Intelsat launch, would be delayed. Vibrations caused chips to fall off the computer circuit board.[33] Teledyne worked around the clock and found a solution to the problem. By means of a different bonding technique, the chips remained securely fastened.

Despite the controversy over the two-burns (which NASA refused to give up) and the concerns over the Teledyne computer chips, the launch went forward. During the first launch attempt on 8 December 1974, a malfunction was detected in the Centaur liquid-hydrogen pump. When the cause of the problem was not resolved during the standard 10-minute delay, engineers postponed the launch. They later discovered that the problem lay with a faulty transducer assembly, which they easily replaced. Two days later, the giant vehicle lifted off.

The first Titan core burn lasted 258 seconds, at which point the stage was shut down and jettisoned. One second later, the Titan second stage engines started. The next major launch event was the jettisoning of the Centaur Standard Shroud (CSS), which occurred 318 seconds after liftoff. At 469 seconds into the launch, the Titan second stage engines shut down. Four seconds later, Centaur separated from Titan. At this point, both German and American engineers collectively held their breaths because this was the exact moment at which the Proof Flight had failed. Eleven seconds later, Centaur passed the critical test when its main engines started. This first burn lasted for 100 seconds. At this time, the Centaur, Delta TE-364, and Helios were traveling at 25,600 feet per second. The main engines then shut down on schedule and Centaur began a nearly 22-minute parking orbit.

Toward the end of the parking orbit, Centaur's thrusters reoriented the vehicle in preparation for the second burn. This burn also occurred on schedule and lasted for 273 seconds, giving the rocket a velocity of 37,400 feet per second. The second main engine cutoff occurred at 36

---

[32] Interview with Joe Nieberding by Virginia Dawson and Mark Bowles, 15 April 1999.

[33] Daniel Shramo, quoted in Craig Covault, "Centaur Computer Trouble May Delay Helios, Intelsat," *Aviation Week & Space Technology* (5 August 1974): 48.

minutes after liftoff, and the engine remained silent for 72 seconds after the end of the second Centaur burn. Then the Delta TE-364, mated to the Helios probe, separated from Centaur. The Delta TE-364 motor started 42 seconds later and burned for 44 seconds. Helios separated after 72 seconds and journeyed alone the rest of the way to the Sun.[34] The successful launch meant a great deal to the Titan-Centaur teams. Writing in *Aviation Week and Space Technology*, Craig Covault said, "Launch of the Helios solar probe from Cape Canaveral last week is as much a success for NASA's Titan Centaur program as it is for the West German team that built the Helios spacecraft."[35]

With Helios safely on a trajectory to the Sun, Centaur began an additional set of experiments to improve future performance. After the Delta TE-364 separated, Centaur began a 1-hour coast period—double the length of the previous successful parking orbit. During this period, Centaur slowly turned itself to point toward Earth so that ground engineers could pick up more data from the rocket. To assist in this procedure, data transmission was switched to two high-gain, narrow-beam antennas. At the end of the 1-hour coast, Centaur began its third engine start in space with a burn of 11 seconds. This was followed by another coast period and included several 180° rolls of the rocket (called thermal maneuvers, which kept one side of the rocket from overheating), a vent sequence, and other thermal control operations. After 3 hours, the Centaur engines had their fourth and final burn, which lasted 47 seconds. This final coast phase lasted almost 27 minutes and included a boost-pump experiment to determine the rocket's capability to recover from a severe cavitation condition (when the pump sucks gas rather than liquid propellant), as well as a hydrogen peroxide depletion experiment. As a result of these post-Helios experiments, Lewis Research Center engineers concluded, "All of the Centaur systems performed satisfactorily, the design parameters for zero gravity coasting were verified, and no significant problems were encountered."[36]

The post-Helios experiments proved that Centaur was capable not only of four-burn missions, but also of vastly extended parking orbits. This capability was accomplished through six key modifications to the basic Centaur design: a new three-layered aluminized Mylar radiation shield that dramatically reduced the hydrogen tank sidewall heating, an improved tank vent control system that enabled the propellant tanks to vent only when necessary, the addition of new "purges" to keep key lines free of liquid, the ability to engage in a 180° thermal roll every 28 minutes to keep a uniform temperature on the rocket, the capability to fire hydrogen peroxide axial thrusters only to collect propellants during a tank vent or engine restart, and, finally, the

[34] K. R. Krebs, "TC-2 Centaur Stage Post Flight Data Review at LeRC," 28 and 29 January 1975, General Dynamics, Box TC-1 TC-2 TC-3 Records, Division Atlas/Centaur Project Office, NASA GRC Records.

[35] Craig Covault, "Helios Launch Verifies Titan Centaur," *Aviation Week and Space Technology* (16 December 1974).

[36] A. B. Yanke, "TC-2 Centaur Stage Post Flight Data Review at LeRC," 28 and 29 January 1975, General Dynamics, Box TC-1 TC-2 TC-3 Records, Division Atlas/Centaur Project Office, NASA GRC Records.

implementation of improved tank-pressurization techniques. These experiments established a major new capability for future missions.[37] Engineers at Lewis predicted that more than 7 hours of coasting would have been possible before tank venting was necessary.

The Helios probe contributed significantly to humankind's knowledge of the Sun. Just over three months after liftoff, the spacecraft reached its closest point to the Sun, 28.7 million miles, on 15 March 1975. Data from Helios were sent back to scientists at the German Control Center in Oberpfaffenhofen, West Germany. All of the scientific instruments functioned normally and sent back information from the closest point to the Sun to which any craft had ever voyaged. The Helios craft produced new information about the solar wind, the emission of a stream of plasma from the Sun, which distorts the magnetic field and produces a shock wave similar to a wind. Study of the solar wind is important because every second, the Sun expels roughly one million tons of ionized particles at great speeds. This wind is propelled throughout the solar system, greatly affecting Earth's environment. Helios voyaged to the outer solar corona, where these particles picked up their final, massive accelerations. Scientists hoped that data gathered at this point would be very revealing about the nature of these particles.

Scientists and engineers on the Viking program also celebrated the success of Helios. Because Mars launch windows open every two years, any launch malfunction would have delayed for two years the planned Mars voyage that was scheduled to take place nine months after Helios. Walter Jakobowski, Viking program manager, said that while watching the Helios launch, "We were holding our breath. That launch vehicle had to work for us. We were very pleased with the overall performance."[38] Ironically, after the launch of the second Viking, a fire on the launch pad threatened to jeopardize the second Helios flight. With help from the Air Force, the launch facility was repaired in time for the launch of Helios 2 (TC-5) on 15 January 1976.

This was the fifth Titan-Centaur launch, with two successful Viking launches occurring between the two Helios launches. The launch vehicle and launch sequence for Helios 2 were nearly identical to those of Helios 1. In April 1976, this probe passed 2 million miles closer to the Sun than Helios 1. As a result of this close passage, the Helios craft was subjected to 10 percent more heat than Helios 1 and continued to operate through temperatures of over 700°F, the melting point of lead. Terry Terhune, Engineering Chief of the Centaur Operations Division at Kennedy Space Center, called Helios 2 a perfect mission with flawless operation of all ten

---

[37] Raymond F. Lacovic, "Centaur Zero Gravity Coast and Engine Restart Demonstration on the TC-2 Extended Mission," Lewis Research Center, October 1975, Box TC-1 TC-2 TC-3 Records, Division Atlas/Centaur Project Office, NASA GRC Records.

[38] Walter Jakobowski, quoted in Victor K. McElheny, "76 Mars Landing Gains With Launching of Titan," *New York Times* (18 December 1974): 15.

experiments (seven German and three American) and absolutely no problems with the any of the spacecraft's systems.[39]

As with the previous voyage, Lewis engineers were able to perform several post-Helios flight maneuvers. Centaur had about 4,000 pounds of fuel remaining after Helios separated. This massive fuel supply enabled Centaur to make an unprecedented five additional engine restarts. These experiments demonstrated a high-altitude, synchronous-orbit injection capability with a parking orbit that lasted over 5 hours between the second coast and the third engine start. Centaur also tested a very short coast of 5 minutes.[40]

The German engineers who originally designed the Helios probes had set a specific definition of what results would constitute a "success" for the mission. They had decided that if the craft returned data from its launch date to the end of the first perihelion passage, plus a mission extension of eighteen months, they would consider it a success. But the unexpected happened. They greatly underestimated the abilities of their own craft. Helios 1 had a lifespan of eleven years, and Helios 2 did not die until 3 March 1990, fourteen years after its initial launch.

The decade-long gathering of data from the Helios probes continues to actively shape scientific knowledge of the Sun. In 1984, at the ten-year anniversary of the program, the Germans invited all of the American scientists who had participated in the project to Germany for an all-expenses-paid trip. American scientist Al Opp summarized the significance of Helios. He said that Helios had "given the scientific world a detailed, close-in view of the Sun over vastly differing solar conditions. It has enabled scientists to observe cosmic rays coming into the solar system from our galaxy, and when combined with results from deep space probes and earth-orbiting satellites, has given a detailed picture of the structure of the solar system and of the characteristics of low energy galactic cosmic rays."[41]

Interestingly, the number of scientists working on the Helios project has increased with time. Although several of the scientists who designed and ran the first experiments are still active in the program, there are many more scientists now studying the Sun with Helios data than were ever originally involved with the project. This expansion was made possible through the Helios international data centers, which provide information to the wider space science community. Rainer Schwenn and Eckart Marsch, two of the scientists working on Helios today, wrote that these missions "shaped our current perception of the Sun and the inner heliosphere."[42] Equally significant was the Helios contribution to the international approach to space science. Schwenn

[39] C. A. Terhune, "Titan/Centaur-5, Helios 2 KSC Field Report," 30 June 1976, Box TC-5, TC-6 Records, Division Atlas/Centaur Project Office, NASA GRC Records.

[40] Raymond Lacovic, "Propellant Management Report for the TC-5 Extended Mission," September 1977, Box TC-5, TC-6 Records, Division Atlas/Centaur Project Office, NASA GRC Records.

[41] Herbert Porsche, ed., 10 Years Helios (Munich: Wenschow Franzis Druck GmbH, 1984), 9.

[42] Rainer Schwenn and Eckart Marsch, eds., Physics of the Inner Heliosphere I: Large-Scale Phenomena (London: Springer-Verlag, 1990), 6.

Titan-Centaur 2 at Complex 41, Cape Canaveral Air Force Station, with Helios 1 as its payload. The rocket lifted off on 10 December 1974 and was the first successful launch of a Titan-Centaur. (NASA KSC-74P-0298)

and Marsch asserted, "It should be recognized in retrospect that the Helios mission at the time of its conception represented an unprecedented challenge to the . . . network facilities available for national and international collaboration."[43] The significance of Helios remains not only the knowledge of the inner solar system it reaped, but also the international cooperation it fostered.

The Search for Extraterrestrial Life: Viking

The two Viking spacecraft launched in 1975 were massive spacecraft—each included an orbiter and robotic lander. So large were these spacecraft—over 7,600 pounds—that they dwarfed the 2,200-pound Mariner 9, the largest planetary craft yet launched by Atlas-Centaur. With four scientific experiments packed into each orbiter and nine more aboard the landers, the Viking missions represented the heaviest science ever undertaken in outer space. Centaur would once again prove its heavy lift capability and make possible an unprecedented soft landing on the surface of another planet.

Everyone involved with the Viking program knew the expense and prestige associated with it. In the late 1960s, NASA had planned to launch Viking on a Saturn V, but after Saturn production was canceled, NASA selected Titan-Centaur for this important mission.[44] In *Beyond the Moon*, Robert Kraemer described the repercussions for NASA should the mission fail. "Viking was so expensive that a mission failure would shake the entire agency down to its toes, and even a modest cost overrun, say, just 10 percent, could cause the cancellation of several smaller science projects," he wrote. "That would generate a science revolt against future planetary missions that could last for years."[45] With Viking's cost nearing $1 billion, engineers at Lewis knew that the future of NASA's space science program was riding on their vehicle.

Mars has been one of the most compelling objects in the sky throughout history. Because of its strange red color, ancient astronomical observers associated it with fire and blood and named it Mars after the Roman god of war. In 1608, Galileo had discovered the surface features of the planet, and in 1659, Christian Huygens made the first sketches of the planet's dark region, now known as Syrtis Major. With the ability to pinpoint a location on the surface, he was able to prove that Mars rotated like Earth. By the late nineteenth century, Giovanni Schiaparelli and Percival Lowell had published theories that Mars was inhabited by intelligent creatures; they used what appeared to be unnaturally straight lines of canals as evidence. But by the twentieth century scientists were becoming increasingly skeptical of finding life on Mars. The Mariner flyby missions in the 1960s and also the Mariner 9 orbiter in the early 1970s had indicated that Mars was uninhabited, yet they had also revealed a planet that was both dynamic and evolving.

---

[43] Ibid.

[44] Interview with Tom Shaw by Virginia Dawson, 9 November 1999.

[45] Robert S. Kraemer, *Beyond the Moon: A Golden Age of Planetary Exploration: 1971–1978* (Washington: Smithsonian Institution Press, 2000), 136.

This new knowledge made Mars a compelling location for further scientific study. The goal of the Viking missions was to learn more about the planet Mars with special emphasis on the search for extraterrestrial life.[46]

The Viking missions were the third and fourth launches with Titan-Centaur. Viking 1 on TC-4 was launched on 20 August 1975, nine days later than originally scheduled. Within 2 minutes of launch on 11 August, one of the forty-eight valves in the thrust vector control system of one of the two solid rocket boosters of the Titan indicated that it was leaking its propellant. The launch was scrubbed because, as Seymour Himmel said, "You don't commit to launch unless you understand what has happened."[47] The errant valve was removed and replaced. It was determined that the source of the leak was the pintle, a conical steel "plug" that acts to seal the valve outlet when fluid is not supposed to flow. Himmel directed Pratt & Whitney to conduct an x-ray examination of the pintle. The x ray revealed a string-like contaminant in the pintle from base to tip. Having ascertained the nature of the failure and that the replacement valve's pintle did not come from the same production batch, Himmel approved the launch.

With the launch rescheduled for 14 August, technicians now discovered that the Viking orbiter's batteries had dropped from 37 to 9 volts. A rotary switch that was supposed to be turned on 7 minutes after liftoff had been mistakenly left in the "on" position for days. To fix this problem, the entire Viking spacecraft had to be removed for detailed troubleshooting. During this time, the Lewis launch team was also in the midst of managing other launches as well. An Atlas-Centaur Intelsat launch was scheduled sixteen days after Viking. "Our focus was intense," Larry Ross recalled. "We were launching a lot of Atlas-Centaurs right in the middle of the Voyager Viking program on Titan-Centaurs, and that was daunting. And the penalty for missing the planetary window is severe. The time and money are extraordinary."[48]

Finally, on 20 August, Titan-Centaur (TC-4), carrying the first Viking spacecraft, lifted off at the opening of its 71-minute launch window.[49] The vehicle functioned flawlessly through the various intricacies of liftoff. Centaur separated from Titan at the 473-second mark. Centaur then started its main engines 11 seconds later, and they burned for 127 seconds. Centaur, still carrying the heavy Viking payload, then coasted in space for 918 seconds. When this parking mode ended, its main engines started a second burn that lasted for 316 seconds, from roughly the 25th to the 30th minute of the launch. At 2,066 seconds after the launch, Viking separated from

[46] Viking Encounter Press Kit, Box TC-1 TC-2 TC-3 Records, Division Atlas/Centaur Project Office, NASA GRC Records.

[47] Interview with Seymour Himmel by Virginia Dawson, 1 March 2000; and Himmel letter to interviewer dated 14 December 2002.

[48] Interview with Larry Ross by Mark Bowles, 29 February 2000.

[49] "TC-4 Centaur Stage Flight Data Preliminary Analysis," 11 September 1975, Box TC-1 TC-2 TC-3 Records, Division Atlas/Centaur Project Office, NASA GRC Records.

Centaur on a perfect trajectory to Mars. Twenty days later, the second Viking (TC-3) was launched 3 minutes before a potentially mission-canceling thunderstorm. This second launch was virtually identical except for longer coast phases. The parking mode lasted for 1,092 seconds, 174 seconds longer than the first Viking, in order to put the spacecraft on an optimal trajectory to Mars.[50]

On 19 June 1976, Viking 1 first entered into Mars orbit. Although every effort was made to put the lander on the surface of the planet in time to commemorate the United States' July 4th bicentennial celebration, a safe landing site could not be found in time. Through examination of the photos taken from orbit, scientists determined that the first site was unsafe. After studying other options, they agreed on the western slope of Chryse Planitia. In July, Viking 1 touched down on the Martian surface, with its sister ship landing in September 1976 at Utopia Planitia.[51] The mission was scheduled to last 90 days, but it did not officially end until November 1982, when the first lander (which was the last of the four craft to stop transmitting) made its final communication.

The contributions of the Viking missions to humankind's knowledge of Mars were made possible by the ability of Centaur to lift the heavy payload containing the hardware needed for the planned scientific experiments.[52] Although the knowledge gained by these missions did not alter the fundamental understanding of the red planet established by the earlier flyby missions, the results were considered significant. Two space scientists wrote, "The Viking mission did not revolutionize our ideas about Mars as Mariner 9 had done, but building on the foundation of its Mariner and Mars precursors it far surpassed them in the variety, quality and quantity of its data because of four factors—the number of spacecraft, the number of experiments, the duration of the mission, and the telemetry rate."[53]

The orbiters and landers returned thousands of images of the planet covering the entire Martian surface during all of its seasons. These photos also included both of the Mars moons. Of all the information returned from the Viking missions, the most controversial type was the

---

[50] "TC-3 Centaur Stage Preliminary Flight Analysis," 26 September 1975, Box TC-1 TC-2 TC-3 Records, Division Atlas/Centaur Project Office, NASA GRC Records.

[51] Robert Godwin, ed., *Mars: The NASA Mission Reports* (Burlington, Ontario, Canada: Apogee Books, 2000), 175–183.

[52] For further information on the Viking missions, see Eric Burgess, *To The Red Planet* (New York: Columbia University Press, 1978); Robert Zubrin, *The Case for Mars* (New York: The Free Press, 1996); Michael H. Carr, *The Surface of Mars* (New Haven: Yale University Press, 1981); Henry S. F. Cooper, Jr., *The Search for Life on Mars: Evolution of an Idea* (New York: Holt, Rinehart, and Winston, 1980); NASA Viking Lander Imaging Team, *The Martian Landscape* (Washington, DC: NASA SP-425 Government Printing Office, 1978); Cary R. Spitzer, ed., *Viking Orbiter Views of Mars* (Washington, DC: NASA SP-441 Government Printing Office, 1980); Mark Washburn, *Mars at Last!* (New York: G.P. Putnam's Sons, 1977); Gerald A. Soffen and Conway W. Snyder, "The First Viking Mission to Mars," *Science* (27 August 1976): 759; Philip J. Sakimoto, "Viking 1 and 2," in *USA in Space*, eds. Frank N. Magill and Russell R. Tobias, vol. 3 (Pasadena, CA: Salem Press, Inc., 1996), 809–817.

[53] Conway W. Snyder and Vassili I. Moroz, "Spacecraft Exploration of Mars," in *Mars*, eds. Hugh H. Kieffer, Bruce M. Jakosky, Conway W. Snyder, and Mildred S. Matthews (Tuscon and London: The University of Arizona Press, 1992), 103.

biological data.[54] The key conclusion came from a gas chromatograph/mass spectrometer experiment that detected no organic material in the soil.

Although the search for life on Mars had been one of the main reasons that the Vikings were designed and funded, the controversy continues. Pictorial evidence acquired from the Viking orbiter resulted in a perplexing mystery about life on Mars and, for some, evidence that it was once home to an intelligent civilization. On the thirty-fifth pass around the planet, as it was looking for a landing site for the second lander, the first Viking orbiter traveled 1,000 miles above the "Cydonia" region. Photos from this region showed a strange collection of what some called "ruins," including, most ominously, a hollow-eyed, human-like face. While NASA vehemently discounted any extraterrestrial intelligence behind these "findings," others claimed that the truth was being covered up to prevent mass hysteria on Earth. According to science writer Richard C. Hoagland, "Either these features on Mars are natural and this investigation is a complete waste of time, or they are artificial and this is one of the most important discoveries of our entire existence on Earth."[55] What did the Viking orbiter really photograph from its vantage point in the Martian sky? Although the debate continues, the Mars Global Surveyor, launched in November 1996, suggests that "wind erosion and some trick of light and shadow probably account for the Sphinx-like face."[56]

Saving the Grand Tour: Voyager

The final two missions for Titan IIIE-Centaur were considered the pinnacle of NASA's planetary efforts of the 1970s. Already the most productive era in the exploration of the solar system, the Voyager flights to Jupiter, Saturn, and beyond promised to yield the most exciting and significant information yet about Earth's largest planetary neighbors. That these probes would not die, but would slowly fade away into the darkness of the galaxy in the hopes of one day encountering intelligent life, also fueled the imagination of popular culture. For example, in the first *Star Trek* movie, the crew of the Enterprise confronts a strange alien intelligence known only as VGER. By film's end, the audience learns that the entity is one of the Voyager probes, which has become self-aware over centuries of seeking knowledge in space, yet which incorrectly names itself because certain letters on the craft have eroded over time. NASA's own words helped fuel some of these fantastic ideas. One JPL press kit stated that "the Voyagers are able to care for themselves and perform long, detailed and complex scientific surveys without continual commanding from the ground."[57] Planetary Programs Director Robert Kraemer said

---

[54] Hugh H. Kieffer, Bruce M. Jakosky, and Conway W. Snyder, "The Planet Mars: From Antiquity to the Present," *Mars*, 15.

[55] Richard C. Hoagland, *The Monuments of Mars: A City on the Edge of Forever* (Berkeley: North Atlantic Books, 1987).

[56] John Noble Wilford, *Mars Beckons: The Mysteries, The Challenges, The Expectations of Our Next Great Adventure in Space* (New York: Alfred A. Knopf, 1990), 111.

[57] "JPL Press Kit, Voyager," Box TC-5, TC-6, TC-7 Records, Division Atlas/Centaur Project Office, NASA GRC Records.

The Centaur stage undergoes tests in the Vertical Integration Building at Kennedy Space Center on 19 October 1976. This Centaur was mated with a Titan for a Voyager launch one year later. (NASA KSC_76PC_0526)

that project scientists "had designed a great deal of artificial intelligence into Voyager's brain."[58] This type of rhetoric, combined with the imagination of science fiction, made it a short leap to conceive of Voyager's becoming self-aware.

Actually, with much less fanfare or fictional speculation, the computer intelligence of the Centaur rocket literally saved Voyager 1 after a nearly fatal error by Titan. Voyager 2 launched first because Voyager 1 traveled on a more efficient trajectory. Launched 16 days after Voyager 2, Voyager 1 was scheduled to beat its sister ship to Saturn. (NASA engineers place more emphasis on arrival date than liftoff order.)

Titan-Centaur (TC-7) looked majestic next to the gantry as it sat poised for the launch of Voyager 2 in August 1977. One day before liftoff, Bruce Murray, the Director of JPL, and his wife stood on the gantry, 160 feet above Pad 41, gazing at the spacecraft. In the distance, lightning illuminated an ominous sky, and they quickly planned their escape route if the storms came closer. They were standing next to the largest United States launch vehicle, loaded with 700 tons of volatile fuel. Murray recalled, "As we gingerly descended the winding stairway of the gantry tower on that glorious Florida afternoon, a quick glimpse of a busy engineer working inside the Centaur surprised me. It was a reminder that unmanned rockets do indeed fly by the skill and dedication of humans—on the ground. The engineer was checking this giant's intricate computer brain . . . . The lightening drifted harmlessly seaward into a darkening sky."[59]

During liftoff into the morning skies off Cape Canaveral in Florida on 20 August 1977— exactly two years to the day after the launch of Viking 1—the Titan stages operated flawlessly. The Centaur Standard Shroud was successfully jettisoned 265 seconds after liftoff, and 203 seconds later, Centaur, carrying the Voyager 2, separated from Titan. The main engines of Centaur started 4 seconds later, burning for a total of 101 seconds. At that point, Centaur coasted into a parking orbit around Earth and began its longest zero-gravity coast during an operational mission. The TC-2 and TC-5 postflight maneuvers proved that Centaur was capable of very long zero-gravity coasts in space. The Voyager flight made use of this capability for a nearly 43-minute coast. Richard Geye of Lewis Research Center noted, "This 'zero-g' mode of coast was selected to provide maximum Centaur performance to the Voyager missions."[60]

At the end of the coast, Centaur realigned itself to place it in the exact position desired by the trajectory engineers for the next phase of the flight. Centaur then began its propellant settling procedure before the second main engine start. This second burn lasted 339 seconds, accelerating Centaur and Voyager beyond Earth-escape velocity conditions. After engine

---

[58] Kraemer, *Beyond the Moon*, 189.

[59] Bruce Murray, *Journey into Space: The First Three Decades of Space Exploration* (New York: W. W. Norton & Company, 1989), 143.

[60] R. P. Geye, "Introduction," "Titan/Centaur D-1T TC-7 Voyager 2 Flight Data Report," Box TC-5, TC-6, TC-7 Records, Division Atlas/Centaur Project Office, NASA GRC Records.

cutoff, Centaur coasted for another 89 seconds. During this time, the computer inside Voyager initiated its separation sequence, and Voyager was deployed from Centaur. Fifteen seconds later, the TE-M-364-4 solid rocket motor (the Propulsion Module) on the Voyager ignited for 45 seconds and gave the spacecraft the additional 6,200-feet-per-second velocity beyond the 46,000- to 48,000-feet-per-second velocity that Voyager needed to inject it into the Jupiter transfer orbit. This Propulsion Module weighed approximately 2,700 pounds and was jettisoned after its engines burned out.

What was left after all of these burn phases was the Voyager Mission Module speeding toward a rendezvous with Jupiter roughly two years in the future. The Mission Module contained all of the instruments, communication and data capability, command and control functions, electrical power, and trajectory-adjustment features that were essential for carrying out the scientific objectives. Engineers based its design on Mariner Mars 1971 and the Viking Orbiter.[61]

While the Voyager 2 spacecraft had suffered some technical problems, Titan-Centaur had performed perfectly. This was not the case for Voyager 1 (launched after Voyager 2) on Labor Day, 5 September 1977. Titan's final-stage engine shut down too soon. Over the control room's radio came the announcement, "Solid booster burnout and separation, Titan core ignition . . . Titan burnout and separation." Those in the room immediately knew that there was a problem because the burnout came much sooner than anticipated. The problem had been caused by a failure of Titan hardware, which choked off the propellant flow, leaving 1,200 pounds of propellant unburned. Only if Centaur could make up for this failure during the trajectory-insertion burn could the mission be saved. Bruce Murray, who was watching from the control room, recalls, "This was serious. Titan had underperformed not propelling Centaur and Voyager fast enough." It was the Centaur computers that recognized the deficiency and began to correct the problem. In awe of the Centaur "brain," Murray said that this was the "brilliant part" because it could "compensate for any shortfall in the propulsion of the earlier stage." Centaur extended its own burn and was able to give Voyager the additional velocity it needed. To the great relief of those in the control room, Centaur had saved the day with only 3.4 seconds of burn time remaining. Murray could only shake his head and say, "Wow that was pretty close."[62] Had this flawed Titan rocket been used on Voyager 2, with its more demanding trajectory, Centaur would not have had enough propellants to perform the corrective maneuver. Voyager 1 missed burning up in Earth's atmosphere only because of the power and capability of the Centaur rocket. Centaur had saved half of the Grand

[61] R. P. Geye, "Voyager Spacecraft," "Titan/Centaur D-1T TC-7 Voyager 2 Flight Data Report," Box TC-5, TC-6, TC-7 Records, Division Atlas/Centaur Project Office, NASA GRC Records.

[62] Murray, *Journey Into Space*, 147.

The Voyager spacecraft as it is encapsulated within its protective shroud at the Spacecraft Assembly and Encapsulation Facility at Kennedy Space Center on 2 August 1977. It would later be mated with Titan-Centaur 7 for launch on 20 August 1977. (NASA KSC_77P_0201)

Tour.[63] This occurrence was never widely publicized because it was the last Titan IIIE-Centaur launch.

Both Voyager craft were safely on schedule with a mission to gather the most detailed and scientifically relevant knowledge ever assembled about Jupiter, Saturn, Uranus, Neptune, and the outer solar system. This was the first so-called "Grand Tour" of the solar system, and the planets were aligned in such a way as to make this possible only once every 176 years. With this exact planetary alignment, the Voyager spacecraft could use planetary assists (a maneuver that uses gravitational attraction like a slingshot to assist in propelling spacecraft between planets) to save fuel and time in reaching the distant destinations. Without this technique, decades would have been added to the overall flight time.

Thirteen days into its mission, Voyager 1 turned its camera back toward home and took the first picture of Earth and the Moon in the same frame. Eighteen months later, Voyager 1 encountered Jupiter, with its closest approach on 5 March 1979. Four months later, Voyager 2 reached the planet. The Voyagers then employed another slingshot maneuver, using Jupiter itself to propel them toward Saturn, altering their elliptical orbit around the Sun. This maneuver officially began the Grand Tour, with a destination beyond the solar system. By the time both reached Saturn, Voyager 1 was nine months ahead of its sister ship.

At this point, the paths of the Voyagers diverged. Voyager 1 sacrificed future planetary encounters for a close examination of Saturn's largest moon, Titan. Voyager 2 continued its planetary tour and made the first-ever encounter of Uranus on 24 January 1986. Three years later, it visited Neptune and its moon Triton, officially ending the Grand Tour and marking the beginning of its interstellar mission. The Voyager missions were a great success, providing the first detailed visual and scientific studies of the Jovian and Saturnian systems, as well as providing the first close encounters with the planets of the outer solar system.[64] Thus the Titan-Centaur launch combination once again played an essential role in the planetary space program, and even when Titan experienced technical difficulties, the power and "intelligence" of Centaur were able to save the mission.

However, despite what should have been a time of great celebration for the Centaur team, the team members faced the termination of the program that they had spent years of their lives

---

[63] Linda J. Horn, "Voyager 1: Jupiter," in *USA in Space*, 823.

[64] For further information on the Voyager spacecraft, see David Morrison, *Voyages to Saturn* (Washington, DC: NASA SP-451 Government Printing Office, 1982); David Morrison and Jane Samz, *Voyage to Jupiter* (Washington, DC: NASA SP-439 Government Printing Office, 1980); Margaret Poynter and Arthur L. Lane, *Voyager: Story of a Space Mission* (New York: Macmillan, 1981); Henry S. Cooper, *Imaging Saturn: The Voyager Flights to Saturn* (New York: Holt, Rinehart, and Winston, 1985); Joel Davis, *The Interplanetary Odyssey of Voyager 2* (New York: Atheneum Publishers, 1987); Eric Burgess, *Far Encounter: The Neptune System* (New York: Columbia University, 1991); Committee on Science, Space, and Technology, *Voyager 2 Flyby of Neptune* (Washington, DC: Government Printing Office, 1990); Rick Gore, "Voyager Views Jupiter's Dazzling Realm," *National Geographic* 157 (January 1980): 2–29; Rick Gore, "Uranus: Voyager Visits a Dark Planet," *National Geographic* 170 (August 1986): 178–195; L. Soderblom, "The Galilean Moons of Jupiter," *Scientific American* 242 (January 1980): 88–100; and William J. Kossman, "The Voyager Program," in *USA in Space*, 819.

Mating Voyager 2 to Titan-Centaur 7 on 5 August 1977. (NASA KSC_77P_0209)

in perfecting. Andy Stofan, reflecting on what he considered to be the shortsightedness of NASA planners, commented, "After the launch of the last Voyager, there were no more Titan-Centaurs ordered, and that was the largest lift capability this country had."[65] Joe Nieberding lamented, "All that effort spent to integrate the Centaur to the Titan, and we launched it only seven times."[66] The concern was not just that Titan-Centaur was no longer going to fly, but that all expendable rockets were to be phased out in favor of the Space Shuttle.

Approved in 1972 during an era of drastically reduced NASA budgets, the new reusable launch system was presented to the Nixon administration and to Congress as a more economical launch system than expendable rockets. Indeed, space science had never figured prominently in the wider context of NASA funding priorities. As historian Roger Launius has pointed out, in the 1960s and 1970s, NASA's extremely productive space science missions were unobtrusively folded into the large Apollo budgets.[67] However, in the 1970s, the huge expense of projected planetary missions like Galileo, Cassini, and Magellan garnered unwanted congressional debate and scrutiny.

[65] Interview with Andrew Stofan by Mark Bowles, 13 April 2000.

[66] Interview with Joe Nieberding by Virginia Dawson and Mark Bowles, 15 April 1999.

[67] Robert Kraemer, Foreword to *Beyond the Moon*, xvi.

Titan-Centaur 7 lifting off from Complex 41 at Cape Canaveral Air Force Station on 20 August 1977 with Voyager 2 as payload. (NASA KSC_77PC_0269_01)

Of the cancellation of Titan-Centaur, Launius wrote, "Because the performance of the shuttle and its then-planned upper stage were less than that of the Titan Centaur, the cancellation decreased the size of the payload that NASA could send to the planets."[68]

Although the advance of technology often leaves well-designed systems in its wake, the larger issue was one of potential over-dependence on a single type of launching system. NASA was essentially forsaking two decades of service by reliable expendable rockets for an unproven system that promised only minor advantages (along with numerous disadvantages) over existing systems. With astronauts to be unnecessarily involved in the future planetary program, significant safety issues arose. From the point of view of mission specialists at JPL, the most unfortunate consequence of the decision was the loss of Centaur, the upper stage upon which they relied to get their heaviest payload into space. Tom Shaw, a JPL engineer, recalled that in the "rush to divest ourselves of expendable launch vehicles and go put all of our eggs in the Shuttle basket," JPL came close to getting an official reprimand from NASA Headquarters for its vociferous objection to the phasing out of the expendable launch vehicles upon which it had relied for the decade's most important scientific missions.[69] But the decision had been made.

---

[68] Ibid., xvii.

[69] Interview with Lutha (Tom) Shaw by Virginia Dawson, 10 November 1999.

Chapter 6

Centaur Reborn

"At stake is the Shuttle's viability as a launch vehicle for the massive new communications satellites being developed for the latter half of the decade— as well as the viability of new NASA plans to send spacecraft into remote parts of the solar system."

—M. Mitchell Waldrop, "Centaur Wars," *Science*, 1982

The rebirth of Centaur in the early 1980s stemmed from a radical idea to redesign the rocket to fit inside the Space Shuttle. Shuttle/Centaur was considered to be the most significant new program at NASA in the early 1980s. Because the Shuttle was limited to flying in low-Earth orbit, it needed an upper stage. Shuttle/Centaur promised to solve this Shuttle limitation. Once in low-Earth orbit (621 miles above Earth), the astronauts could launch Centaur from the Shuttle bay. Shuttle/Centaur was capable of either delivering its payload to upper-Earth orbit or placing it on a trajectory to other planets or the far corners of the solar system. These payloads included communications satellites, Earth-orbiting scientific spacecraft for both NASA and the Department of Defense, and interplanetary probes.[1]

Shuttle/Centaur was hailed as the convergence of piloted and robotic spaceflight and the next step in launch vehicle evolution. In an era of cost cutting and declining space budgets, the combination of the Shuttle with the world's most powerful upper stage promised to give the United States a system of extraordinary power and versatility. Because of these benefits, NASA Associate Administrator Jesse Moore believed that Shuttle/Centaur would become a "very integral, longtime part of the space shuttle program."[2]

Despite the hopes and dreams that surrounded the new rocket combination, it proved extraordinarily controversial throughout the period of development. NASA suspended the program after the *Challenger* disaster in 1986, then officially terminated it without

---

[1] General Dynamics report, "Centaur/Shuttle Integration Study" (Contract NAS 3-16786, 1 August 1973), 2, Drawer 4-C, NASA Glenn Research Center (GRC) Archives.

[2] Robert Locke quotes Jesse W. Moore in "General Dynamics Shows Off New Probe-Launching Rocket," *San Diego Tribune* (14 August 1985).

Shuttle/Centaur ever having flown. Like the original Centaur first conceived in the late 1950s, Shuttle/Centaur was an innovative concept that was yet to be proven. However, unlike in that era, when technical decisions were driven by the Cold War, Shuttle/Centaur was developed during an era with a much lower tolerance for risk.

Today, most people cite concerns over safety as the central reason that Shuttle/Centaur never made the journey into space. But according to the core group of engineers most responsible for designing and building the new Centaur, all of the safety precautions were addressed and all necessary emergency redundancy plans were in place. Although the safety concerns were real, something more powerful was at stake in the decision to abort such an expensive and critical program. A more accurate explanation for why NASA prevented the launch of Shuttle/Centaur was that a devastating human tragedy forced the entire space program into an era of heightened risk sensitivity.

Because Shuttle/Centaur promised to significantly increase its capability to launch key national security missions for the Air Force, the risks associated with liquid hydrogen had initially appeared acceptable. However, détente with Soviet Union and a new Air Force determination not to depend exclusively on the Shuttle for launching its national security payloads made the need for Shuttle/Centaur less urgent. After *Challenger* exploded in January 1986, taking the lives of the entire crew, NASA decided that launching Centaur from within the Shuttle was an unacceptable risk.

The Shuttle/Centaur saga suggests that the subjective, yet changing, tolerance for risk influences technical designs, funding, and management. Everyone involved–politicians, industry contractors, and NASA personnel–is willing to assume more risk when national security is at stake. The Apollo program is an example of a grand dream with significant risk that was unquestioned because it was ideologically important for America to beat the Soviet Union to the Moon. Without this incentive, it is impossible to say whether humans would have flown on top of a stack of rockets with volatile liquid-hydrogen upper stages. By 1986, with competition between the superpowers diminishing and the horrific images of a Shuttle exploding off the Cape's coast shown daily on the news, a decreasing risk tolerance spelled the end for Shuttle/Centaur.

In addition to the changing tolerance for risk, the story of Shuttle/Centaur illustrates the "colliding cultures" between the piloted and robotic space program. Although rivalry between NASA Centers was not unusual, expendable launch vehicles and the piloted space program had operated in distinctly different spheres since the beginning of the Apollo Program. For the first time, the manned and unmanned cultures collided forcefully in the contentious relationship between Lewis Research Center (the home of Centaur) and Johnson Space Center (the home of astronauts and the Space Shuttle). Although Lewis Research Center and General Dynamics had formed a collegial relationship with each other over a period of thirty years, neither had any experience in dealing with the NASA bureaucracy that operated the Shuttle program.

Shuttle/Centaur mission patch. (Glenn Research Center unprocessed photo)

Choosing Centaur for the Shuttle

Although the Shuttle promised a revolution, its technical capabilities left some concerned. Since it could only attain a low-Earth orbit, it needed an upper stage to push some payloads to a geosynchronous transfer orbit—an extremely elliptical orbit with a perigee (closest approach to Earth) of approximately 125 miles and an apogee (farthest distance from Earth) of about 22,300 miles. One logical choice to fill this requirement was Centaur. But how difficult would it be to use the "old" robotic expendable workhorse on a new piloted space vehicle?

Although the liquid-hydrogen upper stages of Saturn V had carried the astronauts to the Moon, Centaur had never been "man-rated." In 1961, the Space Taskforce Group at Langley had considered the possibilities of using the rocket for piloted missions but decided that Centaur was not safe to carry astronauts.[3] Ten years later, testifying before Congress after several failures of Atlas-Centaur, NASA Deputy Administrator George Low said that without

---

[3] James R. Hansen, *Spaceflight Revolution: NASA Langley Research Center from Sputnik to Apollo* (Washington, DC: NASA SP-4308, 1995), 283.

major and costly changes, the vehicle could not be expected to achieve reliabilities much greater than about 90 percent.[4]

However, General Dynamics believed that not only could Centaur satisfy the new upper stage requirements for the Shuttle, but it also could be made safe enough to be carried in the cargo bay. By April 1972, the company had completed a compatibility study assessing safety and performance capabilities for using a liquid-hydrogen rocket with the Space Shuttle. Centaur had already flown thirty-seven missions and had a flight backlog planned through 1979. The report concluded, "The Centaur can be adapted to Shuttle operation with a minimum of modification."[5]

In 1973, the Convair Aerospace Division of General Dynamics issued a more detailed, eight-month "Reusable Centaur Study." This study established the feasibility of modifying the existing Centaur for upper stage launches from the Shuttle. It analyzed twelve key aspects of the program, including such important factors as safety, cost, schedule, weight, reliability, reusability, and flight complexity. The report concluded that on a scale of 0 (low risk) to 10 (high risk), none of these factors were rated higher than 3. The final assessment of the report was that the "Centaur programs are extremely low risk" and that no technology breakthroughs were required to achieve a reusable Centaur.[6]

A heated controversy developed over whether Centaur was the answer for the Shuttle upper stage. Should NASA engineers redesign Centaur as General Dynamics proposed, or should an entirely new upper stage be developed? An author in the journal *Science* reported, "Arcane though it sounds, the issue ignited a free-for-all between feuding congressional committees, the aerospace lobby, the Reagan White House, the Air Force, and NASA, with the latter caught in the middle."[7] The way these issues were resolved would possibly shape much of America's exploration of space for the foreseeable future. These interactions included the growing market for launching large new communication satellites, as well as NASA's capability to send new probes further into the solar system.

In the early 1970s, NASA planned to develop what it called a "space tug" for lifting Shuttle payloads into higher altitude orbits or boosting spacecraft to other planets.[8] In January 1974, NASA awarded a $1.3-million contract to Rocketdyne to design, build, and test liquid-hydrogen

---

[4] George Low, "Review of Recent Launch Failures," statement, Hearings before the Subcommittee on NASA Oversight, 15–17 June 1971, 8.

[5] *Compatibility Study of a Cryogenic Upper Stage With Space Shuttle*, 17 February 1972, Drawer H-D, NASA GRC Archives; and *Compatibility Study of a Cryogenic Upper Stage With Space Shuttle*, Final Report, April 1972, NASA GRC Archives, Drawer H-D.

[6] "Reusable Centaur Study," General Dynamics report, Contract NAS 8-30290, 26 September 1973, 222, 233, Drawer 4-C, NASA GRC Archives.

[7] M. Mitchell Waldrop, "Centaur Wars," *Science* 217 (10 September 1982): 1012–1014.

[8] David Barker, "Economics of the Space Shuttle," *Flight International* (29 August 1974): 244–246.

Artist's rendering of Shuttle/Centaur preparing to launch payload. The bay doors are open, and the payload is about to be released by the Centaur Integrated Support Structure (CISS). (Glenn Research Center unprocessed photo)

and oxygen pumps and gas generators that would eventually be used in the tug.[9] NASA debated the actual design of the tug and whether it would be a modification of an existing upper stage or an entirely new rocket. Four proposals were evaluated including the Air Force's Inertial Upper Stage (IUS), a redesigned wide-body Centaur, Transtage, and the Interim Orbital Transfer Vehicle (IOTV).

Although all four of these stages could lift Shuttle payloads into geosynchronous orbit, each did so with very different levels of capability. A more powerful upper stage rocket permits the launch of heavier, more complex spacecraft that help scientists conduct more sophisticated research. Of all the options, Centaur was the most powerful, enabling spacecraft of up to 13,000 pounds to be sent into orbit. The IOTV, like Centaur, was a cryogenic rocket and was designed to give a similar performance. The next most powerful was the Transtage, which was a storable propellant system capable of lifting spacecraft of up to 8,000 pounds into orbit, followed by the IUS, which had a 5,000-pound lifting capability.

The Department of Defense (DOD) gave the following assessment of each of these rockets. They concluded that the IUS would be able to satisfy most defense needs and, with modifications, could also take on a limited number of basic science missions. The Transtage could satisfy all of the defense needs but did not have the potential for expanding into other types of missions. DOD concluded, "Shuttle payload limits (65,000 pounds) will limit both the IUS and Transtage growth such that these systems can never capture a significant portion of the long term defense needs."[10]

The IOTV and Centaur were both powerful cryogenic rockets that DOD believed would not only handle all existing military needs, but could also provide tremendous potential for more difficult science missions. The problem with the IOTV was that it was a new stage involving a great deal of development complexity, as well as schedule and cost risks. As a result, NASA and the Department of Defense concluded, "The Centaur is the only vehicle capable of meeting near term NASA planetary requirements . . . [and] considerable enhancement of DOD mission capabilities."[11]

However, the development of the IUS continued. This upper stage eventually began to launch smaller spacecraft, with a successful first flight on a Titan 34D booster in October 1982. The operational IUS was 17 feet long and 9 feet in diameter, and it weighed in excess of 32,000 pounds, including 27,000 pounds of solid rocket fuel. But problems occurred on the second flight when the IUS was spring-launched from the Shuttle on mission STS-6. Designed to carry a Tracking and Data Relay Satellite System (TDRSS) satellite into orbit, the IUS rocket motor nozzle accidentally changed its position by 1°, causing the IUS and the satellite to tumble into an incorrect orbit. It took two years to investigate and understand the failure to ensure that it would not happen again.

Although Centaur development appeared promising at this point, the proponents of the IUS mounted a vigorous campaign for their rocket. Initially, several key people from NASA opposed the General Dynamics Centaur plan and agreed with those who sought an alternative to the Centaur upper stage solution. They argued that the IUS, which was designed by the Air Force, was a potentially better rocket. The first stage of the two-stage rocket was capable of launching medium-sized payloads at most. This limitation would be overcome by means of the addition of a second stage for larger payloads with destinations into deeper space. Specifically, the Air Force asked NASA to develop an additional stage that could be used for planetary missions such as a proposed probe to Jupiter called Galileo. NASA made Boeing the prime contractor for developing the IUS.

---

[10] Ibid.

[11] Ibid.

Artist's rendering of the Shuttle/Centaur. The Centaur and its payload have separated from the Shuttle and are preparing to launch after the Shuttle flies away to a safe distance. (Glenn Research Center unprocessed photo)

Success for Boeing meant not only building a technically sound rocket, but also attacking their competition–Centaur and General Dynamics. Boeing made every attempt to demonstrate that the IUS should replace Centaur for use on the Shuttle. Boeing argued that Centaur (they derisively referred to it as the "fat tank" instead of the "wide body") was risky in the area of safety considerations. They also thought it unlikely that Centaur could be ready by the 1985 deadline and claimed that General Dynamics had significantly underestimated the risk associated with reconfiguring the craft. Boeing contended that its rocket offered a much greater chance for overall success and argued that while Centaur required a major new development program, the Boeing alternative was safer because it used solid propellants. The IUS minimized risks for the crew, was mission-flexible, and was Shuttle-compatible. Boeing concluded, "The modified Centaur cannot satisfy the mission needs."[12]

The White House and some members of Congress seemed to agree. The Office of Management and Budget (OMB) was against Centaur because it believed that there were too many modifications required to make the rocket fit in the Shuttle. If these modifications were

---

[12] "Boeing Rebuttal to Shuttle/Centaur," undated, Drawer 4-C, NASA GRC Archives.

to be attempted, OMB also believed that they would be far too costly. Ronnie G. Flippo, a Democratic Congressman from Huntsville, Alabama, agreed with the White House's decision and actively lobbied against Centaur. Flippo was the chairman of the House Subcommittee on Space Science and Applications, which was the committee that authorized all NASA funds. Flippo attacked Centaur in a number of letters to his colleagues. His argument was that Centaur was too expensive (at least $634 million), was of limited usefulness (primarily for just two space missions), and was an example of the faulty sole-source procurement model (with General Dynamics being the prime contractor without competitive bids). Lurking strongly in the background of Flippo's argument was local politics. His Alabama district included Marshall Space Flight Center, which was designated as the Lead Center to manage any future NASA IUS work.[13] However, Centaur did have its supporters in Congress, most notably Bill Lowery, who was a California Republican. But just as local politics influenced Flippo, so too did they play a major role in Lowery's advocacy for Centaur. General Dynamics was located within his San Diego district. Despite Flippo's bias, NASA agreed with him over the protests of Lowery, and they decided to utilize the Air Force-developed IUS.

In his *Journey Into Space*, Bruce Murray argues that this important subcommittee under Flippo "looked at NASA through an Alabama prism."[14] Marshall was desperately looking for ways to regain greater responsibility in the space program. During the Saturn V era in the 1960s, Huntsville, Alabama, was a prime location for all the major aerospace contractors. When the Apollo program ended in the 1970s and the charismatic leader of Marshall Space Flight Center, Wernher von Braun, was stricken with cancer, the Center suddenly faced difficult times. Its rival, Johnson Space Center, was awarded the Space Shuttle program, and the Marshall engineers were faced with taking direction from another NASA Center. As a result, the mood at Marshall became more and more defensive, and through Flippo, they vigorously rejected the Centaur-in-Shuttle idea.

Besides Flippo's advocacy, NASA had other reasons for selecting the IUS as the best choice for the Shuttle. The consensus was that Centaur was too dangerous to fly in the Shuttle along with humans. Centaur was never conceived as a human-rated machine. Its pressure-stabilized tank design saved weight, but it was not necessarily seen as compatible with human flight. There were two main concerns: Centaur's lack of structure reinforcement and the inability of the Shuttle to land safely if Centaur could not be deployed and remained in the cargo bay. Centaur was essentially one continuous tank that had no internal reinforcement to give it added stability. A double-walled bulkhead kept the liquid hydrogen separated from the liquid oxygen. Heat transfer between the two cryogenic liquids was prevented by a phenomenon known as "cryo-

---

[13] M. Mitchell Waldrop, "Centaur Wars," *Science* 217 (10 September 1982): 1012–1014.

[14] Bruce Murray, *Journey Into Space: The First Three Decades of Space Exploration* (New York: W.W. Norton & Company, 1989): 213.

genic pumping" that produced a vacuum between the two walls of the bulkhead when the much colder liquid hydrogen was loaded. The vacuum maintained the liquid hydrogen at -423°F and the liquid oxygen at -297°F.

The human-rated Saturn V's upper stages had a very different design. Each had two separate propellant tanks. The entire structure was reinforced internally by a system of ribs. Like Saturn V, the Shuttle had a sturdy, reinforced structure, and it also carried liquid hydrogen and liquid oxygen. Critics focused on Centaur's pressure-stabilized tank design and feared that somehow the liquid hydrogen and liquid oxygen would prematurely come into contact with each other and explode. Reflecting on the structure, one engineer stated, "You would not be able to sit down and in some rational technical way show that it was dangerous, but intuition and good engineering would dictate that you don't want that feature."[15]

The most challenging of the safety-related issues was what to do if the Shuttle mission was aborted before Centaur was deployed. With human beings involved, the safety issues became a serious concern. What would happen if one of the crew became sick, or the bay doors would not open and Centaur was not deployed? The Shuttle could not land with the hazardous liquid propellants in its cargo hold. The only option was a complicated fuel dump. This was technically feasible, but the process was complex. On Atlas-Centaur or Titan-Centaur, the worst that might happen was that it might explode, resulting in a financial loss. Not only would it be dangerous to try to land with Centaur still in the Shuttle, but also, even with Centaur propellants dumped overboard to lower the weight and the risk, the Shuttle itself might not survive the landing. The Shuttle is a very poor gliding machine because it loses one mile of altitude for every three miles of distance and because its landing gear and braking systems were not designed for the 12 to 15 tons that Centaur would add to its weight.[16] These factors initially made the IUS the upper stage of choice.

However, Boeing was unable to solve key problems with the IUS. Its main limitation was that it was not powerful enough to launch a payload to a distant planet like Jupiter in a direct shot. To get to its destination, the payload would have to perform "mission design tricks" to accomplish a slingshot maneuver around several planets in order to gain a gravity-assisted speed boost. Most engineers considered this approach inelegant and too time-consuming. They did not want to wait additional years for their probes to reach their destinations. Also, new concerns about the cost of the Inertial Upper Stage booster emerged. Robert A. Frosch, upon leaving his post as NASA Administrator, argued that the possibilities of readying the IUS to send an orbiter and probe to Jupiter for its scheduled launch were remote. Frosch decided on 15 January 1981 that the only alternative was Centaur. He argued that by allocating budget resources from 1981 and 1982,

---

[15] Interview with Larry Ross by Mark Bowles, 1 March 2000.

[16] Murray, *Journey Into Space*, 234.

modifications of the Centaur could result in a "powerful combination." He concluded, "No other alternative upper stage is available on a reasonable schedule or with comparable costs. The Shuttle/Centaur would offer both to commercial customers and to national security interests a highly capable launch vehicle with growth potential."[17]

Larry Ross, a Lewis Centaur engineer and project manager, agreed with this assessment and argued that national security interests represented the key reason why the Centaur was chosen over the IUS. Ross said that the main question was that if NASA "married Centaur with the Shuttle for the planetary program, what does the nation end up with?"[18] His answer was that the nation ended up with a very significant capability. Of the two rockets—the Shuttle/Centaur and the IUS—the benefits of the former to the nation far outweighed the benefits of the latter. As a result, Centaur was chosen, both because planetary missions needed the extra boost Centaur could provide and because it better suited missions controlled by the Department of Defense and especially the Air Force. Thus, national security interests may have raised the risk tolerance for the more dangerous rocket.

In 1981, NASA withdrew support for the IUS and instead opted for General Dynamics's redesigned Centaur. A number of technical factors made the Centaur option more attractive. First and most important, Centaur was more powerful. It had the capability of delivering the boost necessary to propel a payload directly to a planet in the deep solar system. Second, Centaur was a gentler rocket. Solid rockets had a harsh initial thrust that had the potential to damage their delicate payload. Liquid rockets generated their thrust more slowly, thus reducing this threat. Finally, a rocket that used liquid propellants had one other tremendous advantage over one that used solid fuel—it could be turned off and on. Once ignited, solid fuel burned until used up, while the liquid was much more controllable.[19] Centaur had already demonstrated the success of this capability with the various two-burn missions for Atlas-Centaur, in addition to the multiple restarts performed in the post-mission experiments for the two Titan-Centaur Helios missions. The only remaining advantage of the IUS over the Centaur was safety. Liquid hydrogen was a dangerous fuel and presented a significant risk to the rocket and the astronauts flying with it. NASA decided to take the chance and go with Centaur.

This decision was met with mixed reviews from the Air Force; on the one hand, it was disappointed in NASA's decision to abandon the IUS in favor of Centaur. On the other hand, the Air Force branch responsible for developing large military payloads was pleased with the additional capabilities that Centaur provided. However, the White House Office of Management

[17] Robert Frosch, "Frosch on Centaur," *Lewis News* (30 January 1981): 1, NASA GRC Archives.

[18] Interview with Larry Ross by Mark Bowles, 1 March 2000.

[19] Waldrop, "Centaur Wars," 1012.

and Budget mounted a more vocal protest and threatened to cut funding. Flippo continued to attack the Centaur decision vigorously, and he launched an intense fight to reverse NASA's Shuttle/Centaur decision when Congress began its next session in February 1981.[20] Eventually, NASA was forced into a compromise, relinquishing its payload role in the International Solar Polar Mission (saving roughly $50 million) in exchange for the right to keep Centaur and the Galileo mission to Jupiter. In August 1982, Congress officially gave its approval to develop the new Centaur high-energy upper stage for the Space Transportation System. Only time would tell whether this decision was the correct one to make.

While the IUS struggled in its early operational years, NASA was considering another critical decision—which NASA Center would take the responsibility for redesigning Centaur. Ultimately, the responsibility for managing the Centaur redesign fell upon the shoulders of NASA Lewis Research Center in Cleveland, Ohio.[21] However, before the decision was made, engineers at Lewis had to withstand a political play by the powerful Johnson and Kennedy Space Centers to keep the contract out of Lewis's hands and give it to Marshall.

Internal Politics: The Push for Marshall

In the early 1980s, Lewis Research Center was a troubled organization. Over the past decade, the laboratory had weathered the closing of its nuclear rocket propulsion program and a devastating Reduction in Force (RIF) that had affected over 700 engineers. At the same time that it was trying to secure a position in the mainstream of NASA—the piloted space program—it was fighting for its very survival. The Heritage Foundation, a conservative Republican think tank, recommended cutting federal spending by eliminating what it considered unnecessary government research and technology. With Lewis Research Center on the Reagan administration's short list of Centers to be closed, the staff mounted an active campaign to save the Center.[22]

Gaining a role in Space Shuttle missions was part of the strategy to reclaim the reputation of this once-proud laboratory. Throughout the 1970s, the engineers and research scientists at Lewis had made important contributions to the Shuttle. They were NASA's experts in propulsion and energy conversion and had performed basic studies on rocket engine turbomachinery, reduced gravity phenomena, chemical energy conversion, and flight simulation testing.[23] They hoped that their greatest challenge lay ahead—redesigning the most powerful robotic rocket to fit inside the piloted Shuttle. The difficulty lay not only in the redesign of the rocket, but also in plans for failure contingencies, such as what would happen if Centaur had to be brought back

---

[20] Murray, *Journey Into Space*, 214.

[21] In 1999, this Center was renamed NASA Glenn Research Center at Lewis Field.

[22] See Virginia Dawson, *Engines and Innovation*, 212–213.

[23] Martin J. Braun, "Technology Transfer: Lewis Contributions to Space Shuttle," *Lewis News* (26 February 1982): 3–4, NASA GRC Archives.

to Earth without having been deployed. For example, how could Centaur's fuel be safely drained? Questions like these were not an issue for other rockets like Redstone, Atlas, or Titan because they were all first stages with no option not to fire.

In 1979, John Yardley, Associate Administrator for Space Transportation Systems, instructed Lewis Research Center to determine the feasibility of integrating Centaur with the Shuttle. By contracting with General Dynamics/Convair, Lewis engineers planned to develop a design that would minimize modifications to both vehicles. The main area of concern at this time was the "fulfillment of safety requirements imposed on all Shuttle cargo as a result of 'man-rating.'"[24] Lewis concluded that a cryogenic upper stage would provide increased mission flexibility for the Shuttle and would meet the unique safety requirements. This positive finding was very important for Lewis. The Space Shuttle threatened the expendable launch program, including its own Centaur program. By proving the feasibility of Shuttle/Centaur, Lewis could maintain its Centaur work and also significantly enhance its involvement in the space program.[25] Implicit in this preliminary work was the belief among Lewis engineers that if Shuttle/Centaur were approved, they would win the management of the Centaur conversion.

Lewis Research Center encountered strong opposition from within NASA. The leadership at Johnson Space Center (JSC), Marshall Space Flight Center (MSFC), and Kennedy Space Center (KSC) all opposed a Lewis-led Centaur program for Shuttle, and they recommended that Marshall should take the lead. In a dramatic display of unity and persuasion, the Directors of these Centers (Chris C. Kraft, Jr., William R. Lucas, and Dick G. Smith, respectively) wrote a confidential joint letter to Alan M. Lovelace, then-acting NASA Administrator, in January 1981. They marked this letter "eyes only," meaning that they did not want their controversial position to leak out to the wider NASA community. Their argument was simple and straightforward— they did not want Lewis to take any leadership position with Centaur. They believed that Marshall represented the best choice. They wrote:

> The Agency's decision to proceed with the development of the Wide Body Centaur as a planetary upper stage in lieu of the Inertial Upper Stage presents a most challenging assignment for the Agency . . . . Recognizing this complexity, it is essential that the Agency make the proper assignment of the management responsibilities for this program . . . . Therefore, we believe, when consideration is given to all relevant factors, it is in the best interest of the program and the Agency that the existing JSC/MSFC/KSC team be responsible for the Wide Body Centaur program.[26]

---

[24] Lawrence J. Ross, "Launch Vehicles," *Lewis News* (4 January 1980): 4, NASA GRC Archives.

[25] Neil Steinberg, "Centaur Star Still Glows Brightly in Agency Plans," *Lewis News* (15 August 1980): 4, NASA GRC Archives.

[26] Chris C. Kraft, Jr., William R. Lucas, and Dick G. Smith to Alan M. Lovelace, 19 January 1981, Larry Ross personal document collection.

Within this team, the directors believed that Marshall should specifically shoulder the burden of managing Centaur for the following reasons. First, they argued that Marshall had a "unique capability" for developing cryogenic rocket engines and propulsion systems. Also, because a significant technical challenge for Shuttle/Centaur was in the area of structures and dynamics, they believed that Marshall had the most experienced team to meet these demands. Second, they argued that the Shuttle was a complicated piloted system and that only their Centers had the "thorough understanding" of the technical and managerial relationships to integrate a new element like Centaur. As a result, the three Directors concluded, "The JSC/MSFC/KSC STS team clearly recognizes the necessity for utilizing the Centaur system as it now exists . . . and is prepared to commit to the discipline required to implement this approach."[27] They also garnered support for Marshall from other important NASA leaders such as John Yardley, who was the head of the Shuttle program at the time.

Despite Marshall's desire to keep its latest recommendation private, some engineers at Lewis Research Center saw a copy of the confidential letter. Larry Ross, Director of Space Flight Systems at Lewis, recalled, "I was outraged. I never showed [the letter] to the team."[28] Lewis engineers believed that because of their experience in managing the Centaur Program, they should be the ones to manage the development of Shuttle/Centaur.

In March 1981, Lewis Director John F. McCarthy wrote his own highly persuasive letter to Lovelace, making the argument for his Center to win the contract. McCarthy listed many reasons why Lewis was the right choice. First, Lewis had led the program to determine the feasibility of modifying Centaur for the Shuttle. This experience would be invaluable in following these recommendations and building the new rocket. Second, Lewis engineers had the greatest Centaur experience among all the NASA Centers. Specifically, their experience with the successful Titan-Centaur program provided a model on which to base the Shuttle/Centaur program. The effort to reconfigure the Atlas-Centaur for Titan included many of the same conversion issues that would arise with Shuttle. Third, Lewis had great experience with mission design and spacecraft integration. With Atlas-Centaur; Titan-Centaur; Earth-orbital missions; lunar missions with Surveyor; and Earth-escape missions like Viking, Voyager, and Pioneer, this experience meant, according to McCarthy, that "no other NASA center approaches the level of expertise in this area that Lewis has attained through the integration of such a wide variety of complex missions." In fact, the average number of years that each Lewis engineer had on the Centaur team was an impressive thirteen. McCarthy concluded, "Our staff uniquely possesses both the knowledge and experience and by virtue of its stability preserves an irreplaceable corporate memory . . . I strongly believe that Lewis continues to be the best choice for Centaur management."[29]

---

[27] Interview with Larry Ross by Mark Bowles, 1 March 2000.

[28] Ibid.

[29] John F. McCarthy, Jr., to Alan M. Lovelace, 25 March 1981, Larry Ross personal document collection.

One month later, in May 1981, Lovelace made his decision. In a letter to William R. Lucas, Director of Marshall, he wrote that he had thoroughly reviewed his proposal for Marshall's control of Centaur and acknowledged the support given to him by both Johnson and Kennedy. He wrote, "Although I recognize each of the considerations which you mentioned as contributing to this recommendation, I have concluded that they are outweighed by another set of factors stemming from the long and continuing management of the Centaur program by the Lewis Research Center." Primarily, these factors included the Lewis experience and their excellent track record with the rocket. It was clear that Lewis was the right choice. The authors of the history of Marshall pointed out that Lewis "had built the Centaur, and had staked out a role in advanced propulsion technology that Marshall could not expect to emulate."[30] Despite the significant competition, Lewis won the right to manage the Shuttle/Centaur program.[31]

Staking the Lewis Reputation on Centaur

Although ecstatic over its victory, the Centaur team knew it was staking its reputation on the success of Shuttle/Centaur. NASA had a strong political commitment to the Shuttle/Centaur program, and any failure at Lewis would lead to "embarrassment in extremis." Larry Ross warned that if the Center "failed in this vital Agency effort, I would despair of us ever again being relied on to do important work for NASA!"[32]

The Center quickly began organizing contractors for the daunting task of integrating Centaur with the Shuttle. Lewis submitted to General Dynamics a request for a proposal to provide materials, personnel, supplies, and services to create two versions of the new wide-body Centaur—one for NASA and the other for the Air Force.[33] By June, Lewis had awarded four contracts totaling $7,483,000 to four different companies: General Dynamics to develop two modified Centaur vehicles; Teledyne Industries for a digital computer and several remote multiplexor units; Honeywell, Inc., for part of the automatic navigation and guidance system; and Pratt & Whitney for four RL10A-3-3A rocket engines. All of these contracts were geared toward two key missions—the upcoming Galileo probe to Jupiter and the Ulysses (formerly the

---

[30] Andrew J. Dunar and Stephen P. Waring, *Power to Explore: A History of the Marshall Space Flight Center, 1960–1990* (Washington, DC: NASA SP-4313, 1999), 138.

[31] Alan Lovelace to William R. Lucas, 27 May 1981, Larry Ross personal document collection.

[32] Larry Ross to Chief Counsel, 7 December 1983, Box 2142, Space Transportation Engineering Division (STED), Project Management Office, Folder General Correspondence, NASA GRC Archives.

[33] "Technical Evaluation of GDC Proposal, Shuttle/Centaur G Development and Production," 15 December 1983, Box 2142, Division STED Project Management Office, Folder Centaur G Program, NASA GRC Archives.

International Solar Polar) mission to the Sun.[34] Despite the start of work and the flow of funds, uncertainties hovered over the project for over a year and constantly threatened to suspend the project. Would the funding and support for Shuttle/Centaur continue? These questions could not yet be answered.

Adding to the uncertainty was a change in leadership at Lewis Research Center. Director Dr. John F. McCarthy, Jr., announced that he planned to retire in July 1982 and return to MIT as a professor of aeronautics and astronautics, where he had previously headed the Center for Space Research.[35] Andrew Stofan then became director of Lewis. His experience in launch vehicles would serve him well in the coming years. Stofan began his career as a research engineer at Lewis in 1958. Four years later, he moved to the first Centaur project office, where he was part of the Propellant Systems team. He continued to move up the launch vehicle ranks at Lewis, heading the Titan-Centaur Project Office. He served as director of Launch Vehicles from 1974 to 1978. He was later promoted to Associate Administrator at NASA Headquarters; then he returned to Lewis as McCarthy's successor in 1982.[36]

Stofan's arrival in his new position at Lewis coincided with the major announcement for the Shuttle/Centaur program. President Ronald Reagan signed an Urgent Supplemental Appropriations Bill that allocated $80 million for the design, development, and procurement of the Centaur upper stage. This was a dramatic increase over 1981 funding for this project, which had amounted to only $20 million. Stofan said that this new funding put "Lewis directly into the mainstream of the Shuttle program."[37] In a memo to Lewis employees, he expressed his enthusiasm over the fact that Congress had decided to continue this project. Stofan believed that this congressional decision represented the overturning of the last potential threat to the program and that they could at last "expect to carry on from this point without the specter of precipitous cancellation." Stofan considered the new connection with the high-profile Shuttle program "a major step forward for Lewis."[38] Lewis budgets began to increase substantially for the

---

[34] Jim Kukowski, "Contracts Awarded for Shuttle Launched Upper Stage," 2 June 1981, NASA Headquarters Press Release; Linda Peterson, "NASA Lewis Awards $7.5 Million in Contracts for Shuttle/Centaur Development," 2 June 1981, Lewis Research Center (LeRC) Press Release, GRC Public Affairs Archives; "Four Contracts Awarded for Modifying Centaur," *Lewis News* (2 July 1981): 3, NASA GRC Public Affairs Archives.

[35] Mary Fitzpatrick, "Lewis Director to Leave NASA to Return to MIT," 2 March 1982, NASA Headquarters Press Release, NASA GRC Public Affairs Archives.

[36] William O'Donnell, "Stofan to Head Lewis Research Center," 19 March 1982, NASA Headquarters Press Release, NASA GRC Public Affairs Archives.

[37] Paul T. Bohn, "NASA Lewis Gets Added Funding of $107.4 Million," 27 July 1982, LeRC Press Release, GRC Public Affairs Archives; "Center in Mainstream of Shuttle Program," *Lewis News* (30 July 1982): 1; "Additional Funds Resurrect Important Programs;" Lewis News (13 August 1982): 1, 6, NASA GRC Archives.

[38] Andrew J. Stofan to Lewis employees, 17 September 1982, Box 2142, Division STED Project Management Office, Folder General Correspondence, NASA GRC Archives.

project. In 1983, the Center received $133 million. In 1984, this sum increased to $159 million, and it increased again to $188 million in 1985.[39]

Not everyone within Lewis was as enthusiastic over the Center's new management responsibilities. A large Lewis contingent believed that this work did not fall within the mission of the Center to do basic research. Joe Nieberding recalled, "To those folks, Centaur was a sink for a lot of resources, potentially distracting management attention and NASA resources from the 'real' Lewis charter: to do research. The research side saw little direct involvement for them in Centaur, and they were mostly correct."[40]

Increased funding for Centaur allowed for an increase in the Lewis workforce for the first time in twenty years. In 1983, 190 new engineers were hired. Stofan said that this hiring reflected "a strong Lewis effort to reinvigorate its professional staff, and the permission to so expand represents a vote of confidence from Washington in the Center's future."[41] The main area of activity for these new hires was in helping to transform the Centaur launch vehicle into a Shuttle upper stage.

Lewis formed the Shuttle/Centaur Project Office to serve as the central point of integration and management for the program. This office developed all Centaur project objectives and evaluated the progress of these goals. These goals included defining Centaur redesign requirements, creating the new vehicle design, determining how best to integrate it with the Shuttle, and finally developing and producing the new Centaur. This office also controlled the budget for the program and was responsible for scheduling and coordinating all contractors and internal work related to the reconfiguration.[42] This project office had extensive relationships with NASA Headquarters; Johnson Space Center; Kennedy Space Center; and outside organizations, contractors, and other government agencies including the Department of Defense and the European Space Agency.[43]

While this office relied upon the expertise of all of its workers, there were several unique individuals who helped to shape its future. William H. "Red" Robbins took over as head of the Shuttle/Centaur project office. He was a longtime NASA engineer with experience in nuclear propulsion, and this was his first exposure to Centaur. Robbins served as project manager and was responsible for the administrative functions including budgets and schedules. Steven V. Szabo headed the engineering division and, in 1983, became the new chief of the Lewis Space

---

[39] "Director Reports 'Very Good' 1984 Budget," *Lewis News* (11 February 1983): 1; "Stofan Charts Labs Future," GRC Archives, *Lewis News* (4 November 1983): 1, NASA GRC Archives.

[40] Joe Nieberding, correspondence with authors, 7 November 2001.

[41] Paul T. Bohn, "190 New Engineer Hires Increasing Lewis Work Force for First Time in Two Decades," 15 July 1983, Lewis Press Release, NASA GRC Public Affairs Archives.

[42] Larry Ross, "Space Flight Systems," *Lewis News* (4 February 1983): 2, NASA GRC Archives.

[43] Shuttle/Centaur Project Office, Functional Statements from the NASA Organizational Manual, 1 October 1983, Box 2143, Division STED Project Management Office, Folder Shuttle/Centaur Organization–LeRC, NASA GRC Archives.

Transportation Engineering Division. Another longtime Centaur engineer, Szabo began work in the original Centaur project office in 1963. In his new position, he led a one-hundred-member team responsible for keeping Atlas-Centaur operational and for implementing the new Shuttle/Centaur program.

As the program evolved, Steve Szabo became the critical cog in the development of Shuttle/Centaur. As chief of the transportation and engineering division, he was directly responsible for managing the myriad of technical activities required to integrate the Centaur with the Shuttle. His responsibility included all propulsion, pressurization, structural, electrical guidance and control, and telemetry hardware systems, as well as the flight software systems and the structural dynamics, controls, trajectories, and thermal analysis activities. Within the Shuttle/Centaur project office, Edwin Muckley headed the Mission Integration Office. He had the complicated job of integrating each of the payloads assigned to Centaur. Frank Spurlock managed trajectory mission design for Shuttle/Centaur.[44] Joe Nieberding took charge of the Shuttle/Centaur group within the Space Transportation Engineering Division responsible for development and management of all software and analyses. Spurlock and Nieberding hired many of the young engineers who later became the division chiefs, branch chiefs, and program managers.[45] With an experienced team of veterans, supplemented by young, bright scientists and engineers, Shuttle/Centaur moved forward.

Building the New Horse

By the early 1980s, the Centaur boasted fifty-three successful payload launches into space. Centaur was heralded as reliable. It had sent payloads to destinations like the Moon, Mercury, Venus, Mars, Jupiter, and Saturn; it had also successfully placed many communications satellites into geosynchronous orbit and several heavy telescopes into Earth orbit. Shuttle/Centaur promised Centaur a continued and long life of planetary exploration. However, reconfiguring the "premiere upper stage" rocket to launch from inside the Shuttle was not an easy task.[46] Working from a limited budget and timeframe, the Shuttle/Centaur redesign team struggled to keep changes to a minimum because each one increased the cost, decreased the reliability, and lengthened the development period.

The "first major milestone event" in building the Shuttle/Centaur was a meeting held at General Dynamics in San Diego. On 30 August 1982, NASA and all Shuttle/Centaur contractors met to discuss a System Requirements Review (SRR) for the entire project. Harry O. Eastman III from the General Dynamics Shuttle/Centaur project office wrote, "The purpose of the SRR is to

---

[44] Charles Owens, "Five Space Flight Systems Engineers Promoted at NASA Lewis," 27 July 1983, LeRC Press Release, NASA GRC Public Affairs Archives.

[45] Interview with Joe Nieberding by Virginia Dawson and Mark Bowles, 15 April 1999.

[46] W. F. Rector III and Don Charhut, "Centaur for the Shuttle Era," May 1984, Box 2142, Division STED Project Management Office, Folder Technical Paper Approvals by PM, NASA GRC Archives.

assure that the Shuttle/Centaur project system level requirements are adequately stated, and to determine that the elements can indeed meet the program requirements."[47] Glynn Lunney, the manager of the Space Shuttle program at Johnson Space Center, wrote that this meeting would be of vital importance: "It is important that requirements that cannot be met, are of insufficient detail, or are in error, be addressed at this time."[48] With the completion of this milestone, NASA could commence building the "new horse."

There were similarities between the new Centaur and the old version. The propulsion system was virtually the same. The avionics system was still mounted on the forward equipment module, and it still included the Teledyne Digital Computer Unit. This was a 16K, 24-bit computer that controlled guidance and navigation by integrating acceleration data to determine position and velocity. The Centaur pressure-stabilized tank was still made of the same thin stainless steel that had served it well for over twenty years. However, the diameter of the hydrogen tank expanded to accommodate additional fuel for longer planetary voyages. The tank also required insulation. The forward bulkhead used a two-layer foam blanket and a three-layer radiation shield.

Despite the similarity of Shuttle/Centaur to Atlas-Centaur, there were some significant technical changes. Specifically, there were two main redesigns required for the Shuttle to carry the Centaur. First, Centaur and its payload had to fit into the 60-foot cargo bay. Engineers developed two different configurations (called versions G and G-prime). The Centaur G was 20 feet long and was capable of supporting a 40-foot payload to accommodate Department of Defense requirements. By reducing propellant weight to 29,000 pounds, the G version could carry additional weight. It was designed specifically for missions to place satellites into geostationary orbit. The United States Air Force funded half of the initial $269-million expense required to design and develop the Centaur G for use with the Shuttle.[49] The other configuration, Centaur G-prime, was nearly 30 feet long and capable of handling payloads of up to 30 feet.[50] It was optimized for planetary missions, and the payload space was smaller due to the necessary increase in propellant weight (45,000 lb). NASA funded the G-prime configuration.[51] Both versions would fill the entire cargo bay, making it necessary to devote the entire Shuttle mission to launching them. Overall, the two

---

[47] Harry O. Eastman III, 30 August 1982, unprocessed Shuttle/Centaur records, NASA GRC Records.

[48] Glynn S. Lunney, 12 August 1982, unprocessed Shuttle/Centaur records, NASA GRC Records.

[49] NASA/DOD Memorandum of Agreement, November 1982, Box 2141, NASA GRC Records. After the cancellation of Shuttle/Centaur, Centaur G became the upper stage for the Air Force Titan program.

[50] Jeweline H. Richardson, "The Centaur G-Prime: Meeting Mission Needs Today for Tomorrow's Space Environment," May 1983, Box 2142, Division STED Project Management Office, Folder Technical Paper Approvals by PM. See also "NASA Lewis Research Center/USAF Space Division Agreement for the Management of the Shuttle/Centaur Program 10/25/82," Box 2141, NASA GRC Archives.

[51] W. F. Rector III and David Charhut, "Centaur for the Shuttle Era," May 1984, Box 2142, Division STED Project Management Office, Folder Technical Paper Approvals by PM, NASA GRC Records.

versions were very similar, with 80 percent of the design detail being common.[52] The second and most important new technical development was the mechanism needed to support Centaur in the Shuttle and deploy it from the bay.[53] The new deployment device was called the Centaur Integrated Support Structure (CISS), a 15-foot-diameter aluminum structure that attached the rocket to the Shuttle. This device was of central importance because it enabled the Centaur to fly in the Shuttle with a limited number of design modifications. The CISS was located within the cargo bay on rotating support structures. It was responsible for all of the mechanical, electrical, and fluid interfaces between Centaur and the Shuttle, as well as most of the safety precautions for the rocket.[54] The CISS was also fully reusable for ten flights.[55] The development of the CISS was considered one of the "more extensive and challenging tasks" of the entire Centaur modification project.[56]

Along with these two main technical developments, there were hundreds, if not thousands, of smaller integration issues that engineers had to address. A small sample of these tasks included internal thermal integration of the three machines (Shuttle, Centaur, and payload) during launch, deployment, and landing; avionics integration interfaces to monitor tank pressure, power, and computer software between the Shuttle and Centaur; and propulsion integration, which detailed how the propellants would be loaded and also dumped in case of an emergency abort. Countless personnel and general support issues also complicated the list of technical needs. A business-management system was required to control accounting and contract issues and a project office was set up to manage the entire integration process.[57] Estimates from Johnson Space Center indicated that, in total, there were between 500 and 1,500 separate items up for review in the integration process.[58]

Once all of the integration procedures were worked out, the Shuttle/Centaur would be ready for its first flight. The flight operation and the interactions between the Shuttle, Centaur,

---

[52] "Technical Evaluation of GDC Proposal, Shuttle/Centaur G Development and Production," 15 December 1983, Box 2142, Division STED Project Management Office, Folder Centaur G Program, NASA GRC Records.

[53] Harold Hahn (General Dynamics), "A New Addition to the Space transportation System," July 1985, Box 2142, Division STED Project Management Office, Folder Technical Paper Approvals by PM, NASA GRC Records.

[54] Omer F. Spurlock (aerospace engineer, Lewis Research Center), "Shuttle/Centaur—More Capability for the 1980s," Box 2142, Division STED Project Management Office, Folder Technical Paper Approvals by PM, NASA GRC Records. This article is the best technical overview of the entire Shuttle/Centaur.

[55] R. Wood and W. Tang, "Shuttle/Centaur Prestressed Composite Spherical Gas Storage Tank," July 1985, Box 2142, Division STED Project Management Office, Folder Technical Paper Approvals by PM, NASA GRC Records.

[56] Larry Ross, "Space," Lewis News (31 December 1981): 3, NASA GRC Archives.

[57] "Centaur Integration Task Summary," 16 September 1983, unprocessed Shuttle/Centaur records, NASA GRC Records.

[58] Harry O. Eastman, 17 September 1981, Glenn Research Center, unprocessed Shuttle/Centaur records, NASA GRC Records.

and CISS worked in the following way. The Shuttle launch was itself a standard flight. During the launch, engineers at Houston planned to monitor the Centaur through a Tracking and Data Relay Satellite (TDRS) via telemetry links to the ground. Once the Shuttle began its orbit and a predeployment check, the bay doors opened. The rocket was then rotated 45° out of the orbiter bay by the CISS to a launch-ready position twenty minutes before it was to be released. Special springs on the CISS (twelve compressed coil springs called a Super*Zip separation ring) enabled the Shuttle crew to eject the Centaur from the deployment adapter into free flight at 1 foot per second.[59] The Centaur then coasted for 45 minutes prior to its main burn, allowing the astronauts to take the Shuttle a safe distance away. For planetary missions, only a single burn was necessary. Once at the required velocity, the final maneuver was to separate the spacecraft from Centaur. After release, Centaur was capable of maneuvering so that it would not interfere with the flight of the spacecraft and would also prevent planetary impact.

Safety was always a key concern with Shuttle/Centaur. Howard Bonesteel, Manufacturing Director of the Space Systems Division at General Dynamics, remarked that "the big difference is the larger fuel tank for the liquid-hydrogen fuel, and a lot of additional safety systems because it's being carried on a manned spacecraft."[60] One of the major functions of the CISS was to address potential problems with the fluids and avionics systems. Increased safety meant "an ability to tolerate multiple failures before some unwanted consequences take place."[61] Special redundancy features were installed to comply with stringent safety considerations. The engineers believed that these would significantly reduce the dangers of losing liquid-hydrogen fuel. To minimize the risk, a unique propellant fill, drain, and venting system was installed on two of the Shuttles designated to carry Centaur.[62] This system enabled the Shuttle crew to dump the dangerous propellants in case of an emergency, allowing the Shuttle, Centaur, and its payload to return to the ground safely. The benefits of Shuttle/Centaur seemed to outweigh the risks. Larger spacecraft could take longer voyages into the solar system, and much heavier Department of Defense satellites would be able to reach geosynchronous orbits.[63] In short, NASA was aware of the risks from the very beginning but decided that the risk was worth the scientific and national defense rewards.

---

[59] R. E. Martin (from General Dynamics Space Systems Division), "Effects of transient Propellant Dynamics on Deployment of Large Liquid Stages in Zero-Gravity with Application to Shuttle/Centaur," delivered at the 37th International Astronautical Congress, October 1986, Innsbruck, Austria, Box 2142, Division STED Project Management Office, Folder Technical Paper Approvals by PM, NASA GRC Archives. (The separation system was built by Lockheed.)

[60] Robert Locke, quoted in "General Dynamics Shows Off New Probe-Launching Rocket," *San Diego Tribune*, 14 August 1985.

[61] Interview with Larry Ross by Mark Bowles, 1 March 2000.

[62] "Centaur for STS," 8 January 1986, Box 2142, Division STED Project Management Office, Folder General Correspondence, NASA GRC Records.

[63] AP Press release, "General Dynamics," 14 August 1985, Box 2141, NASA GRC Archives.

However, one important question always loomed over the enterprise: was NASA placing too much of its hope for success on the Shuttle and the ideal of reusability? Launch systems like Titan-Centaur were taken out of production because they were expendable vehicles. In the new era of reusability, these old rockets did not satisfy the new vision of spaceflight. However, if the Shuttle's development was halted before completion or the Shuttle was unsuccessful, how would other payloads, such as robotic planetary missions and satellites, make the voyage into space? In 1981, one reporter for the *New York Times* speculated that a Shuttle problem would confirm the "worst fears of those who had criticized the agency for its failure to allow for backup systems to be kept in production while awaiting the shuttle."[64] Little did this reporter realize how prophetic his words were. In five years, before the near launch of the new Centaur, a tragic Shuttle failure left NASA without a launch vehicle.

---

[64] John Noble Wilford, "U.S. to Discontinue New Rocket System," *New York Times* (18 January 1981).

Chapter 7

Eclipsed by Tragedy

*"It was considered to be probably one of the most hazardous projects that
NASA had ever attempted to fly . . . . The chief of the Shuttle office at the
time was John Young and he called Shuttle/Centaur 'Death Star.'"*

−Rick Hauck, NASA Astronaut

*"I'm convinced to this day we would have made the launch window in
May of '86, but it was a sprint to the finish. It was like the racehorse that
overtakes you at the end. Had not the Shuttle accident occurred, we would
have been ready for launch."*

−Marty Winkler, General Dynamics

With the infrastructure in place and the redesigns underway, Shuttle/Centaur was ready to take on a real mission. In the early 1980s, it was designated to launch two of the most significant explorations of the solar system—one probe to Jupiter and the other to the Sun. But in preparing for these payloads, significant and tragic problems arose. The first was in managing Centaur. Project managers and engineers had to overcome significant time constraints because both of these missions were scheduled to launch in May 1986. This complexity was compounded by the problems of making a rocket safe enough to ride with humans and convincing the NASA community that the risks were minimized. The second obstacle was one from which, ultimately, the Shuttle/Centaur could not recover—the tragedy of the *Challenger* explosion in January 1986. In the wake of this disaster, the liquid-hydrogen propellant that coursed through Centaur's veins appeared too risky, and ultimately the Shuttle/Centaur was grounded, eclipsed by a tragedy not of its own making.

The Shuttle tragedy resurrected questions concerning the decision to end the expendable launch program. The ill-fated *Challenger* was scheduled to deliver a Tracking and Data Relay Satellite (TDRS) to a geostationary orbit. Journalist and historian William Burrows pointed out how tragically unnecessary it was for seven people to lose their lives "trying to get a satellite into orbit that could have been sent there on an expendable."[1] Centaur, with either an Atlas or Titan

---

[1] William E. Burrows, *This New Ocean: The Story of the First Space Age* (New York: Random House, 1998), 556.

booster, could have delivered this payload without the risk to human life. Both the risks NASA accepted and those it rejected shaped the future of the Centaur program more than the actual technical success and capability of the rocket itself.

To the Sun and Jupiter

The destinations for the first two Shuttle/Centaur launches were both extremely important in terms of the scientific knowledge they would uncover about our solar system. The Sun and Jupiter, icons of mystery in the sky for as long as humanity has turned its gaze to the heavens, would relinquish some of their secrets to the probes to be delivered by Centaur. Coupled with this importance was a significantly high level of mission complexity for the first Shuttle/Centaur launch system. With only a six-day launch window between the two launches, beginning 16 May 1986, there was no margin for error.[2]

In the 1980s, project Galileo was among NASA's few planetary missions in production.[3] Named for the seventeenth-century Italian scientist who discovered Jupiter's moons, the project had its genesis at JPL in the late 1970s. Its objective was to send two spacecraft to explore the Jovian planetary system. The Galileo team was frustrated by the difficulty in securing a ride for their probe. They went back and forth, redesigning their probe for the IUS, then for Centaur. As late as 1982, this decision changed again when the NASA budget included funding for Galileo but not for Centaur. The IUS again emerged as the launch vehicle.[4] When Centaur was returned to funding status, NASA again abandoned the IUS in favor of a faster ride for the Galileo probe. Initially, the Galileo team was pleased with this decision, although they never suspected that in just four years they would again have to search for an alternate ride.

Equipped with an orbiter and a probe, the Galileo mission would provide unprecedented observations of the fifth planet as well as the four Galilean satellites. It was also scheduled to perform a flyby of the asteroid 29 Amphitrite.[5] This was a large, 200-kilometer

---

[2] John R. Casani to William H. Robbins, 9 April 1985, Box 2143, Division STED Project Management Office, Folder Galileo Correspondence, NASA GRC Archives.

[3] For further information on the Galileo probe, see C. M. Yeates, ed., et al., *Galileo: Exploration of Jupiter's System* (Washington, DC: NASA Government Printing Office, 1985); David Leonard, "Free Fall to Jupiter," *New Scientist* 147 (July 1995): 26–29; C. T. Russell, ed., *The Galileo Mission* (Boston: Kluwer Academic, 1992); Monish R. Chatterjee, "Galileo: Jupiter," *USA in Space*, eds. Frank N. Magill and Russell R. Tobias, vol. 3 (Pasadena, CA: Salem Press, Inc., 1996), 191–194.

[4] John R. Casani to Galileo Review Board, 6 January 1982, Box 2143, Division STED Project Management Office, Folder Galileo Correspondence, NASA GRC Archives.

[5] M. K. Winkler to W. H. Robbins, 23 January 1985, Box 2143, Division STED Project Management Office, Folder Galileo Correspondence, NASA GRC Archives.

main-belt asteroid that was the most interesting of several encounter possibilities for Galileo. However, the asteroid flyby was only considered an optional scientific objective. After launch, engineers would assess the health of the spacecraft to determine whether this option would be exercised. This mission would not only gather scientific knowledge, but would also "significantly enhance U.S. scientific prestige."[6] Once Galileo reached its final destination, the orbiter would become the sixth spacecraft to visit the planet and the first to be placed into orbit around Jupiter.

While orbiting, it would gather scientific data concerning Jupiter's atmosphere and the physical attributes of the satellites. The probe also promised to be the first device to enter the atmosphere of a planet more distant than Mars.[7] Six months after release from the orbiter, it would descend toward Jupiter and record information on the chemical composition of the atmosphere, the structure and physical dynamics of the magnetosphere, and the chemical composition of the Jovian satellites.[8] Scientists hoped that this knowledge would provide greater understanding of the evolution of the solar system, the origin of life, the weather system on Earth, and additional information about Jupiter itself.[9]

In November 1982, Lewis Research Center signed a Memorandum of Agreement with JPL for the Galileo project. Lewis maintained the Shuttle/Centaur project office; JPL held the Galileo project office. The two Centers maintained close contact to ensure that the new Centaur could meet all the specifications necessary for a successful flight to Jupiter. JPL was responsible for managing the entire mission, extending from design concept to mission completion. To Lewis fell "all responsibilities necessary to integrate the Galileo spacecraft with Centaur and the Space Transportation System."[10] This work included coordinating with Johnson Space Center (JSC) and ensuring that there was engineering compatibility and risk management between the Shuttle and Galileo.

The second Shuttle/Centaur mission, also scheduled to launch in May 1986, was a joint venture with the European Space Agency (ESA). Originally called the International Solar Polar

[6] James M. Beggs to Jamie L. Whitten, 24 December 1984, Box 2143, Division STED Project Management Office, Folder Galileo Correspondence, NASA GRC Archives; Jim Kukowski, "NASA May Fly by Asteroid with Galileo Spacecraft," NASA Headquarters Press Release, 27 December 1984, NASA GRC Public Affairs Archives.

[7] Charles Redmond, Peter Waller, "Test of Jupiter Probe Spacecraft a Success," NASA Headquarters Press Release, 11 August 1983, GRC Public Affairs Archives.

[8] "Payload Integration Plan Space Transportation System and Galileo Mission," 24 August 1983, Box 2143, Division STED Project Management Office, Folder Galileo Correspondence, NASA GRC Archives.

[9] "Environmental Impact Statement for Project Galileo," May 1985, Drawer H-D, NASA GRC Archives.

[10] Andrew Stofan and Lew Allen, "Memorandum of Agreement Between the Lewis Research Center and the Jet Propulsion Laboratory for the Galileo Project," November 1982, Box 2142, Division STED Project Management Office, Folder Galileo, NASA GRC Archives.

Mission, the destination for this spacecraft was the Sun and the exploration of the solar environment. It was renamed Ulysses in 1984 to allude both to Homer's hero and to Dante's desire to explore "an uninhabited world behind the Sun."[11] Ironically, in journeying to the Sun, the probe would first travel to Jupiter (the same destination as Galileo) and then use the gravitational mass of the planet to catapult it out of the ecliptic plane (the plane in which all planets orbit the Sun). It was necessary to fly out of this plane, with the associated high energy supplied by the Jupiter gravity-assist, to enable passage over the solar poles. Ulysses was not designed for a close solar approach. In fact, engineers liked to joke that the closest Ulysses ever got to the Sun was when it was sitting on the launch pad in Florida. Nevertheless, this mission would mark the first observations of the solar poles of the Sun. Scientists hoped to uncover knowledge about the solar wind, the heliosphere's magnetic field, the interplanetary magnetic field, cosmic rays, and cosmic dust.[12]

Preliminary plans were also being made for Shuttle/Centaur to launch a third mission, the Venus Radar Mapper, whose name was later changed to Magellan.[13] Lewis Research Center hosted the initial mission integration panel meeting on 8 November 1983.[14] For this mission, scheduled for launch in April 1988, a variety of different upper stage vehicles were examined as launch alternatives. These included the Orbital Sciences Corporation TOS/AMS, Astrotech International's Delta Transfer Stage, and Boeing's IUS.[15] NASA assessed the viability of all these alternatives and considered trajectory, weights, schedule, and risk factors. The evaluators decided that Shuttle/Centaur was the best alternative and sent notice to the competitors that "NASA has concluded that your system is not capable of meeting the VRM [Magellan] mission requirements without imposing unacceptable technical and schedule risks on the program."[16] Once again, the technical merits of Shuttle/Centaur

[11] For further information on the Ulysses probe, see Craig Covault, "European Ulysses Fired to Jupiter, Sun as *Discovery* Returns to Space," *Aviation Week and Space Technology* 133 (15 October 1990): 22; Michael Mecham, "After Long Delay, Ulysses Mission Begins 5-Year Voyage to Expand Solar Data Base," *Aviation Week and Space Technology* (22 October 1990): 111; David H. Hathaway, "Journey to the Heart of the Sun," *Astronomy* 23 (1995): 38; Gordon A. Parker, "Ulysses: Solar-Polar Mission," in *USA in Space*, 785–788.

[12] Jim Kukowski, "Ulysses New Name for International Solar Polar Mission," NASA Headquarters Press Release, 10 September 1984, NASA GRC Public Affairs Archives.

[13] Magellan Mission Design Panel Charter, 7 February 1986, Box 2143, Division STED Project Management Office, Folder VRM Mission, NASA GRC Archives.

[14] Edwin T. Muckley, "Action Items from the First Venus Radar Mapper Shuttle/Centaur Integration Meetings," 8 November 1983, NASA Archives, Box 2143, Division STED Project Management Office, Folder VRM Mission.

[15] Jesse W. Moore to David W. Thompson (Orbital Sciences Corporation CEO), 18 June 1984; Moore to Leonard Rabb (Astrotech International Corporation Director), 22 June 1984; Moore to H. N. Stuverude (Boeing Aerospace Company Vice President), 18 July 1984, Box 2143, Division STED Project Management Office, Folder VRM Mission, NASA Archives.

[16] Jesse W. Moore to Robert J. Goss (Astrotech Space Corporation), 15 August 1984, Box 2143, Division STED Project Management Office, Folder VRM Mission, NASA GRC Records.

had beaten out competition and won the right to launch another prestigious planetary mission.

The Department of Defense (DOD) also began scheduling missions for Shuttle/Centaur.[17] In 1984, it devised a launch plan for Shuttle/Centaur to put the Milstar satellite into orbit.[18] This military communication satellite was designed to operate under the worst conditions, including jamming, interception, and nuclear attack.[19] Plans to use Shuttle/Centaur for this launch were complex. It proved difficult because the DOD required Centaur to perform beyond its designed capability. Meyer Reshotko wrote, "This mission has some important requirements which exceed the Centaur G baseline specifications. Each of these requirements has been requested by the Air Force."[20] For example, the Milstar required a direct attachment to Centaur with explosive bolts used to separate the two. Reshotko said that in order for this to occur, the Centaur team had not only to redesign the satellite interface, but also to begin testing to determine what effects the separation shock would have. Further requirements also included a broader Payload Integration Plan for all DOD payloads destined for launch on Shuttle/Centaur. These were all classified missions; the underlying security clause in each signed agreement between the Air Force and NASA stated, "There will be no public release of information pertaining to payload."[21] This classification made the job of integration far more difficult. For example, if engineers at Lewis wanted to make a telephone call to General Dynamics about the project, they had to go to a "safe" building with secure communications lines.

But again, some asked the question, "Why not send the deep-space planetary missions with an expendable launch vehicle such as Titan-Centaur?" This launch vehicle had proven that it was capable of delivering very heavy scientific missions into the deep solar system. The prowess of Atlas-Centaur as a launch vehicle for communications satellites was unquestioned. In addition to disdain for all things expendable in a new culture that valued reusability, another factor weighed against Titan-Centaur—its military connection. Marty Winkler said, "They [NASA] decided on Shuttle/Centaur which arguably may have been the wrong decision, but that was the decision they took. The reason I believe they did not

[17] Shuttle/Centaur Memorandum of Agreement, approved by Colonel Charles H. MacNevin and William H. Robbins, 17 May 1984, NASA GRC Records.

[18] "Launch Base Test Plan for Milstar Processing at ELS," NASA GRC Records, Box S/C Mission Integration, Division LVPO.

[19] Donald H. Martin, *Communication Satellites*, 4th edition (El Segundo, California: The Aerospace Press, 2000), 208.

[20] Meyer Reshotko to Steve Szabo, "Summary of Out of Scope Mission Unique Requirements," Box S/C Mission Integration, Division LVPO, NASA GRC Records.

[21] "Shuttle/Payload Standard Integration Plan for DOD Deployable/Retrievable-Type Payloads," May 1985, Box S/C Mission Integration, Division LVPO, NASA GRC Records.

want to go with the Titan is that the Air Force owned the Titan, and [NASA] would have to go through the Air Force to do that. Any relationship between the Air Force and NASA, historically, on launch vehicle matters, had been like oil and water."[22] NASA would still work closely with the Air Force on their military communication satellites, but with Shuttle/Centaur, NASA was at least able to limit Air Force involvement to the Centaur G configuration. Whatever the merits of the decision, the Shuttle/Centaur was going into development. The responsibility fell to Lewis Research Center to pull off what many still considered impossible.

Managing Centaur

With the mission objectives established, the main task became managing Centaur so that it could become launch-ready for May 1986. Significant funding for the project did not come until 1983, and the first planetary launches were scheduled for mid-1986. Larry Ross said, "It was a very substantial development program. When we sold it, I don't think we emphasized how much substance there was to the development. I think we said, it's a Centaur, we have them all over the place, we will just make one that will fit in the cargo bay. It was a relatively straightforward technical job, but managerially it was extraordinarily demanding."[23]

The management relations among the various institutions involved in the project were complex. Although the responsibility for Centaur G was jointly shared between NASA and the Air Force, the project management lead fell upon the shoulders of Lewis Research Center, while the Department of Defense and NASA Headquarters provided funding and overall direction for Shuttle/Centaur. The Air Force assigned Major (later Colonel) William Files as the Air Force Deputy Shuttle/Centaur Project Manager and assigned a team of half a dozen other key individuals to become part of the project office in Cleveland. Important interaction and communication were also maintained with other NASA Centers, including Johnson Space Center for the Space Shuttle and Kennedy Space Center for the cargo project office. Mission management included the Jet Propulsion Laboratory for the Galileo project, the Air Force for DOD projects like Milstar, and the European Space Agency for the Ulysses project. The final area of interaction for Lewis was with their contractors. These included General Dynamics for the Centaur vehicle and the CISS, Teledyne for the computer, Honeywell for the guidance system, and Pratt & Whitney for the engines. William H. Robbins, manager of the Shuttle/Centaur Project Office, and Vernon Weyers, deputy manager of the Lewis Shuttle/Centaur Project Office,

---

[22] Interview with Marty Winkler by Virginia Dawson, 21 March 2001.

[23] Interview with Larry Ross by Mark Bowles, 1 March 2000.

called this "complex management arrangement" a challenge that was a "dynamic and demanding assignment for all involved personnel."[24]

Lewis project leaders sought out exceptionally qualified individuals and gave them the support necessary for achieving their goals. To bring together all these individuals from various locations and backgrounds required the development of a team environment. It was hoped that this sense of teamwork would unify the project members and direct their focus upon one key goal—integrating Centaur with the Shuttle. This goal was extremely challenging due to the inflexible time schedule and the number of "firsts" that needed to be accomplished. Ross argued, "The bottom line is that we have a very tough job to do in a very short time and we must succeed."[25]

To help cultivate a team atmosphere, Ross dug deep into what he called his "senior manager's bag of tricks." What emerged were trusted managerial and motivation techniques to concentrate team attention on May 1986, a date three years in the future. Ross said that time was the "single most important aspect of the motivating theme."[26] The schedule was critical for Shuttle/Centaur because the Jupiter launch had only a twenty-one-day launch window. If this launch window were missed, the entire mission would have to be delayed over a year.[27] The cost resulting from such a delay was estimated at about $50 million.

Ross relied on a motivational symbol to convey the urgency, focus, and sense of teamwork required. He wrote, "My campaign centers on a graphic which is specially designed to visually portray the challenge and evoke an active sense of commitment to it."[28] The image itself was an ancient centaur, half-man, half-horse, emerging from the Shuttle, rotating backwards and aiming his arrow into the sky. Ross insisted that the symbol become ubiquitous throughout the various laboratories and offices so that all who worked on the project were visually aware of their goal. Even contractors from industry agreed to display the symbol prominently in their work environments. Ross confessed, "If I am to succeed in this, all team members from our own [Lewis] engineers to mechanics in the contractor's shop will not be able to go through a workday without seeing the symbol and being reminded, and inspired, by the challenge and commitment it represents."[29]

Ross also emblazoned the symbol across a variety of project memorabilia. Campaign-type buttons were printed, along with drink coasters and notepads. To reinforce the importance of

---

[24] William H. Robbins and Vernon Weyers, "Shuttle/Centaur Project Stages Challenge," *Lewis News* (22 March 1985): 1, GRC Archives.

[25] Larry Ross correspondence, Box 2142, Division STED Project Management Office, NASA Archives.

[26] Ibid.

[27] Vernon J. Weyers to Kevin L. Wohlevar, Cray Research, Inc., 16 August 1985, Box 2141, Division STED Project Management Office, Folder General Correspondence, NASA Archives.

[28] Ibid.

[29] Ibid.

meeting all deadlines, the team members received a special Centaur pocket calendar. Unlike most twelve-month calendars, this one covered an unusual twenty-eight-month duration. It represented the months from January 1984 to May 1986, ending with the acceptable window of the Galileo launch. William H. Robbins printed his personal reminder on the first page. He wrote, "It is not possible to overstate the importance of fulfilling our commitment to this challenge . . . . This special calendar is offered as a way of keeping the schedule critical nature of what you do prominent in your daily planning."[30] For the front cover, he combined Ross's symbol with the trite but effective catchphrase from the popular movie *Rocky*: "Go for it!"

Optimism remained strong in 1985 for both the Galileo and Ulysses missions. Fabrication for the G-prime version was well underway, and the structural testing was being completed at the Shuttle/Centaur test facility in Sycamore Canyon, California. Slightly behind schedule were the avionics box and fluid system testing, but this delay did not threaten a major postponement of the program.[31] However, despite this technical and managerial success, the safety concerns would not go away. While it was thought that the safety systems would remove all doubt about the dangers of using the highly explosive fuel, it was exactly this problem that would start a bitter inter-Center conflict at NASA and cast the future of Shuttle/Centaur into doubt.

Centers in Conflict

One of the most significant factors slowing the development of Centaur was the number of interfaces that Lewis handled on a daily basis. The interfaces for Shuttle/Centaur included mission management, the Air Force, customers, systems management at NASA Headquarters, element and cargo management at other NASA Centers, and finally the industrial contractors. Engineers knew that the number of interfaces they encountered during a job increased the overall complexity by geometric proportions. This complexity was further compounded because most of these interfaces were "authoritative," meaning that they had the power to influence or to stop the progress of the overall mission. Marty Winkler, who took over as head of the Shuttle/Centaur program at General Dynamics, recalled how difficult this program was. He said, "In the beginning of 1983 I took over the Shuttle/Centaur program, and spent an extraordinary long and painful three years—very, very pressure filled. There were 70-hour work weeks for three years. I was not a pleasant person to be around. And a lot of it was caused by the internecine warfare [between] the NASA Centers."[32] Specifically, Winkler said, "The people at Johnson Space Center hated the idea. They fought it all the way. It was miserable."

---

[30] Ibid.

[31] Harold Robbins and Vernon Weyers, "Shuttle/Centaur Project Stages Challenge," *Lewis News* (22 March 1985): 1, GRC Archives.

[32] Interview with Marty Winkler by Virginia Dawson, 21 March 2001.

Of all the relationships, Lewis engineers believed that the most "difficult and dominant" one was with Johnson Space Center. As Larry Ross lamented, "Johnson never wanted this program."[33] Red Robbins agreed, "There was an attitude problem at Johnson. They didn't want to fly Centaur. They didn't want a big hydrogen tank sitting behind the astronauts because they thought the risk was too great."[34] Many of the astronauts who were based at Johnson Space Center felt the same way. General Dynamics aerospace engineer Edward Bock said, "The astronauts never became comfortable with the vehicle . . . . There were astronauts that just flat out said, I will not fly the Shuttle with that thing in the payload bay."[35] Tom Shaw, who managed launch vehicle integration at JPL, also commented on the reluctance of astronauts to fly with Centaur. He said, "The people that sit up there in the cockpit looked back over their shoulder and listened to the gurgling and the rumbling of the vent valve opening and closing, and had some reservations. The astronaut office did not feel very kindly towards the use of that big liquid-hydrogen/liquid-oxygen stage in the Shuttle bay."[36] But as was often pointed out, the Shuttle itself was powered with liquid hydrogen and oxygen. Shaw noted that the astronauts were "sitting astride twenty-five Centaurs' worth of liquid oxygen and liquid hydrogen in that external tank. So it's just a question of whether it's between their legs or behind their back is the way I view it."[37]

Why was the Johnson-Lewis relationship strained? The crux of the problem was the understandably cautious attitude Johnson had toward its astronauts. As a Center, Johnson essentially owned the human space program. They were the ones who took the awesome responsibility of risking human life every time they sent a mission into space. Because of this responsibility, crew safety was the first priority. When they faced the need to put a dangerous rocket into their Shuttle, they became highly cautious and conservative. Furthermore, when the decision was made to allow Lewis Research Center to manage the building of this new rocket, and not Marshall as Johnson had recommended, their reluctance increased further. It was Marshall that had always been a member of the manned spaceflight team. Lewis was always on the outside. And when the decision was made to put Lewis squarely in the middle of what was the most significant new space program in the early 1980s, the manned space establishment did not welcome them with open arms.

Johnson Space Center and Lewis Research Center attempted to work together as a team to make Shuttle/Centaur flight-ready, and there were some signs that the earlier disagreements would be put behind them. There was some positive interaction between the two

---

[33] Interview with Larry Ross by Mark Bowles, 1 March 2000.

[34] Interview with William H. Robbins by Mark Bowles, 21 March 2000.

[35] Interview with Edward Bock by Virginia Dawson, 22 March 2001.

[36] Interview with Tom Shaw by Virginia Dawson, 9 November 1999.

[37] Ibid.

Centers. As the project got underway, Andrew Stofan wrote, "cooperation between both centers continues to be excellent."[38] However, under the strain of the short timeframe and high technical complexity of the Shuttle/Centaur program, this relationship was to be severely tested. Specifically, the Centers came into conflict over what was, predictably, the most contentious issue—safety. The dangers inherent in allowing a human crew to carry a liquid-hydrogen rocket into space were significant. Although the safety of the crew and Shuttle was the number-one priority for both Centers, there was often disagreement over the tradeoffs between safety and mission success. Each Center had a different perception of risk.

The first question was how to integrate Centaur with the Shuttle. Would it be considered an "element" or a "payload?" Elements were large, conceptual chunks of a launch vehicle, like the Solid Rocket Boosters, the External Tank, and the orbiter itself. They were the responsibility of Johnson Space Center, where all the safety, reliability, and quality-assurance requirements were developed and managed. A 1981 Memorandum of Agreement between Johnson and Lewis identified Centaur as a "Level III Space Shuttle element."[39]

Initially, Lewis engineers and managers believed that it was in their best interest to reclassify the Centaur as a payload because there was less bureaucracy and documentation associated with this status. With Centaur classified as an element, Johnson would have had more direct control of the project and the budget, and Lewis engineers would have had to take direct orders from them. Classifying Centaur as a payload allowed Lewis to remain more autonomous; the payload handbook guided the specifications. During the monthly safety reviews, Johnson's payload people merely determined if the Lewis work was up to their specifications. With time always a critical factor, Lewis engineers reasoned that the less energy spent in navigating Johnson bureaucracy, the more time they would have for ensuring that Centaur was flight-ready. Red Robbins, head of the Shuttle/Centaur Program at Lewis, argued, "I had so much trouble with JSC the last thing I wanted them to do is muck around in my budget, so I picked the payload option, and it probably can be argued forever whether it was better to be payload or element."[40] Though initially pleased with the new integration arrangements, he later regretted this decision, as did other engineers.[41] Nieberding concluded, "I really believe that had we been an element from the beginning, and worked with the Johnson engineering staff side by side for four years so they could get some confidence in our vehicle, as opposed to treating it as a payload . . . I think we

---

[38] Andrew Stofan correspondence, Box 2142, Division STED Project Management Office, NASA GRC Records.

[39] "Shuttle/Centaur Project Plan," Drawer H-D, NASA GRC Archives.

[40] Interview with William H. Robbins by Mark Bowles, 21 March 2000.

[41] William H. Robbins to Johnson Space Center, 16 August 1983, Box 2142; "Mission Operations Directorate Systems Division Operational Role for Centaur Missions," 27 March 1984, NASA Archives, Box 2142, Division STED Project Management Office, Folder Johnson Correspondence 1984; Glynn S. Lunney to Lewis Research Center, 15 June 1984, Box 2142, Division STED Project Management Office, Folder Johnson Correspondence 1984, NASA GRC Records.

would have had a lot more success."[42] Edward Bock, a member of the Shuttle/Centaur program office at General Dynamics between 1983 and 1986, said that the Shuttle payload criteria applied to satellites and spacecraft that were relatively inert while they were in the cargo bay of the Shuttle:

> When these rules were applied to a high energy upper stage such as Centaur, they resulted in design solutions that were asinine in their complexity and cost. The resulting reliability of these complex design solutions was poor. As these designs matured, this became obvious to all parties involved, and criteria compromises had to be reached to provide rational design solutions. This cost time and money, and the resulting compromises were more akin to "element" design criteria. These "payload" criteria could not have been reasonably applied to any of the other Shuttle "elements," which were designed to different standards. Centaur was an element of the Space System, and should have been designed that way from the start.[43]

After Centaur was redesignated as a payload in 1983, responsibility for performance and reliability for Centaur and CISS shifted to Lewis, with Johnson maintaining the power to override any Lewis decision. Johnson Space Shuttle Manager Glynn S. Lunney (who himself had worked at Lewis Research Center until 1958) explained that this decision was made because "the Centaur project has demonstrated reliability and performance history and the imposition of Shuttle [element] requirements beyond those required to assure interface compatibility and safety is neither cost effective nor consistent with the agreed responsibilities."[44]

With payload status, Centaur was treated like any other piece of cargo, but, in fact, it was a "living, breathing rocket, full of hydrogen."[45] Because this payload was going to be launched from within the Shuttle, it imposed a new reality upon Johnson payload-integration experts. When deployed, the payload influenced the trajectory of the craft and the amount of propellant Centaur required to adjust its course. Johnson Space Center engineers were responsible for this deployment decision, but Lewis had to negotiate with the Johnson payload experts. Frustrations and problems quickly mounted.

Over the next few years, the correspondence between Robbins and Lunney revealed the conflict that surrounded Shuttle/Centaur. Although Robbins initially indicated to Lunney

---

[42] Interview with Joe Nieberding by Virginia Dawson and Mark Bowles, 15 April 1999.

[43] Edward Bock communication to authors, 11 March 2002.

[44] Glynn S. Lunney to NASA Headquarters, 14 April 1983, Box 2142, Division STED Project Management Office, Folder Johnson Correspondence 1983, NASA GRC Archives; Henry C. Dethloff, *Suddenly Tomorrow Came . . . A History of the Johnson Space Center* (Washington, DC: NASA SP-4307, 1993), 278.

[45] Interview with Joe Nieberding by Virginia Dawson and Mark Bowles, 15 April 1999.

that he was pleased with Johnson Space Center internal roles and responsibilities and the plan to organize the entire integration, the relationship soon deteriorated.[46] Robbins always maintained his great respect and admiration for Lunney, whom he described as a "no nonsense, let's get the job done without the politics kind of guy."[47] But these feelings never extended to the rest of Lunney's team at Johnson.

Complicating the payload-versus-element question was the decision by Johnson Space Center to add further requirements for Shuttle/Centaur that went beyond previous Shuttle payloads. One of the most onerous was a fluid systems interface and the special require- ments for dumping the liquid-hydrogen propellant if the mission had to be aborted. In late 1984, Johnson released a Centaur Safety Waiver (WACR-1A) that requested a redesign of the fill, drain, and dump vent system. The problem it wanted addressed was the placement of the liquid-hydrogen and liquid-oxygen vent line inlets below the propellant liquid level in case of a mission abort. The result, according to Lunney, was "potentially catastrophic to the Orbiter and Crew."[48]

Lewis engineers were greatly angered by this new request and believed these safety waivers to be the "fundamental problem all the way through the Shuttle/Centaur program."[49] As early as October 1982, just as the project was commencing, during a Shuttle/Centaur meeting at Johnson, Lunney and others "expressed concern about the safety aspects of combining the Centaur project with the Space Shuttle program."[50] At that meeting, Robbins gave a report to ease the concerns about safety at Johnson. Again and again throughout the project, the safety issue came to the forefront of discussions. Thus, for the latest safety waiver in 1984 concerning the redesign of the propellant-dumping system, Robbins wrote to Lunney that although this request was "technically feasible" and might have been a "reasonably straightforward design task a year ago," now it would have a serious impact on the program.[51] Robbins contended that the redesign would delay Centaur by eight weeks, which jeopardized the fixed launch window of the next mission. Robbins concluded, "I am

---

[46] William H. Robbins to Glynn S. Lunney, 24 November 1982, Box 2142, Division STED Project Management Office, Folder Johnson Correspondence 1982, NASA GRC Records.

[47] Interview with William H. Robbins by Mark Bowles, 21 March 2000.

[48] Glynn S. Lunney to William H. Robbins, 10 October 1984, Box 2142, Division STED Project Management Office, Folder Johnson Correspondence 1984, NASA GRC Records.

[49] Interview with William H. Robbins by Mark Bowles, 21 March 2000.

[50] Minutes for Shuttle/Centaur Board, 28 September 1982, from Melvin E. Dell, concurrence of G. S. Lunney, Glenn Research Center, unprocessed Shuttle/Centaur records, NASA GRC Records.

[51] William H. Robbins to Glynn S. Lunney, 25 October 1984, Box 2142, Division STED Project Management Office, Folder Johnson Correspondence 1984, NASA GRC Records.

not in a position to add risk to the mission launch schedule by accepting and implementing this new design requirement."[52]

Lewis engineers also believed that engineers at Johnson were unresponsive to their needs. The attempt to transition Centaur to a piloted rating was complex, and the engineers required a vast amount of knowledge about the Shuttle. Lewis engineers often felt as if their requests were ignored. As Larry Ross said, "They almost deliberately debilitated the program by not being responsive."[53] Although this lack of responsiveness was partly due to the strained relationship, it was also caused by the fact that this was an extremely difficult time at Johnson. The effort to prepare the early Shuttles for flight was immediate and enormous, and often the requests from engineers working on a payload not scheduled until 1986 took a back seat. Faced with a lack of information, Lewis engineers relied on their judgment to solve a problem. Often, when the Lewis engineers brought their solutions up for review at Johnson, they were rejected. The result was frequent requests for design changes throughout the project.

One such disagreement concerned vehicle testing. Johnson wanted to perform their harness testing on the vehicle itself, while Lewis wanted to test it before installation was complete. Robbins argued, "Our rationale for this position is based on over twenty years of experience on Centaur where this method has been successfully proven and used."[54] Robbins believed that performing the tests after installation would incur an additional cost of $500,000 and delay the project by two weeks. He was not willing to jeopardize the mission by failing to meet the schedule.

In February 1985, Lunney articulated a problem he saw developing between the two Centers.[55] He argued that the Lewis team believed that the problem stemmed from the changing flight requirements at JSC. Others attacked Lewis engineers for their tardiness in identifying problems with the rocket. Lunney proposed a solution of a one-week dedicated effort between the Centers to iron out their differences. The idea was accepted, and Larry Ross wrote, "It's a good idea, and I'm sure it will go a long way in achieving closure on the nagging STS/Centaur safety issues."[56]

---

[52] William H. Robbins to Glynn S. Lunney, 25 October 1984, Box 2142, Division STED Project Management Office, Folder Johnson Correspondence 1984, NASA GRC Records.

[53] Interview with Larry Ross by Mark Bowles, 1 March 2000.

[54] William H. Robbins to Glynn S. Lunney, 7 December 1984, Box 2142, Division STED Project Management Office, Folder Johnson Correspondence 1984, NASA GRC Records.

[55] Glynn S. Lunney to Lewis Research Center, 7 February 1985, Box 2142, Division STED Project Management Office, Folder Johnson Correspondence 1985, NASA GRC Records.

[56] Larry J. Ross to Glynn S. Lunney, 21 February 1985, Box 2142, Division STED Project Management Office, Folder Johnson Correspondence 1985, NASA GRC Records.

The result was the development of a "Centaur flight decisions philosophy" in January 1986. This was a set of mission actions by which the Centaur would be jettisoned from the Shuttle if it encountered any critical system failures such as loss of power from the Shuttle to Centaur, any propellant tank leaks, or any leaks in the helium tanks. Detaching Centaur in these (and other) cases would help preserve the lives of the crew and protect the costly Shuttle itself.[57] These safety issues were still unresolved when NASA experienced one of its most tragic and devastating catastrophes in its history, and one that would spell the end of the Shuttle/Centaur program: the loss of *Challenger* in January 1986. To provide a better understanding of the events leading up to this calamitous event, the following account will trace the yearlong countdown to ready Shuttle/Centaur for launch. Built in San Diego, it left the hands of the contractor and journeyed from west to east to the launch pad in Florida, where it sat the day of the *Challenger* disaster.

T Minus One Year: The Countdown Begins

In May 1985, one year before the first launch of Shuttle/Centaur, the astronaut crews were named for the flight. *Challenger* was the designated Shuttle vehicle to launch the Ulysses probe. The commander was to be Frederick (Rick) H. Hauck, on his third flight after previously piloting a mission in 1983 and commanding another in 1984. His crew included pilot Roy D. Bridges and mission specialists David C. Hilmers and J. Mike Lounge.[58] Pilots were second in command and had the responsibility of controlling and operating the Shuttle. The mission specialists coordinated all Shuttle onboard operations, experiments, and spacewalks.

That same month, Johnson Space Center announced the *Atlantis* crew that would launch the Galileo probe. The commanding officer would be David M. Walker, a captain in the Navy who had flown the Shuttle the year before. The pilot was to be Ronald J. Grabe, a lieutenant colonel in the Air Force. The two mission specialists were James "Ox" Van Hoften and Norman E. Thagard.[59] These were high-profile missions, and everyone expected "the assigned crew members to become highly visible as soon as their scheduled flights are completed."[60]

---

[57] Arnold D. Aldrich, Manager, National Space Transportation System, Johnson Space Center, 13 January 1986, Box 2142, Division STED Project Management Office, Folder Johnson Correspondence 1986, NASA GRC Records.

[58] Charles Redmond and Steve Nesbitt, "NASA Names Astronaut Crews for Ulysses, Galileo Missions," NASA Headquarters Press Release, 31 May 1985, NASA GRC Public Affairs Archives.

[59] Charles Redmond and Steve Nesbitt, "NASA Names Two Space Shuttle Crews," NASA Headquarters Press Release, 19 September 1985, NASA GRC Public Affairs Archives.

[60] Ken Nus, General Dynamics Space Program, general memo, 4 June 1985, Box 2142, Division STED Project Management Office, Folder Johnson Correspondence 1985, NASA GRC Records.

Shuttle/Centaur rollout at General Dynamics, August 1985. (Courtesy of the San Diego Aerospace Museum.)

Rick Hauck was a key member of the team because he was the Astronaut Office project officer for Shuttle/Centaur integration. Hauck was a distinguished pilot and commander. After earning a master's degree in nuclear engineering from MIT, he became a U.S. Navy test pilot. In 1978, NASA began training him as an astronaut for the Shuttle. He served as pilot on STS-7 in June 1983 and commander of STS-51A in November 1984. Throughout the preparation of Shuttle/Centaur, the astronaut crews were heavily involved in many of the decisions regarding the launch, particularly software and procedural development. Because of the risk factors associated with Shuttle/Centaur, Jesse Moore, head of human spaceflight, frequently invited Hauck and Walker to attend senior management meetings where key development issues were discussed. While it was unusual to have astronauts attend these meetings, Moore felt that it was essential for the astronauts to be involved as much as possible. Hauck said, "There was a tremendous amount of focus within the crew on working major issues that probably wasn't typical of many of the other Shuttle missions."[61] These issues included concerns surrounding pushing the Shuttle engines to the 109-percent power level (the previous high was 104 percent). Another problem was that the Shuttle had to orbit at the lowest altitude possible, at 105 miles, because of its limited lift capability. Finally, there were safety issues that the crew had to understand for an abort. If the Shuttle had to return to Earth with Centaur still inside, the center of gravity for the spacecraft would be further aft than ever before, making landing particularly difficult.

Kennedy Space Center was busy making plans for the Galileo and Ulysses launches. The two Shuttles would be housed at Complex 39 on launch pads A and B. To complicate matters, there were only six days scheduled between the launches, with each launch having only a 1-hour window. Officials at Kennedy felt pressure from both the media and the public to duplicate the success of the Viking and Voyager missions and to add to the dramatic images and scientific data these spacecraft returned. A Kennedy briefing warned, "Remember the public's response to . . . pictures of Saturn and Jupiter. They will be looking for more. They and the world will be watching."[62] Lewis Research Center also had the pressure of getting two Centaurs ready for launch. With ten months to go before liftoff, they began their much publicized "Centaur Countdown." They reported that while "much remains to be done, all activities support the crucial schedule."[63]

When scientists and engineers at General Dynamics triumphantly unveiled Shuttle/Centaur (SC-1) in San Diego in August 1985, such excitement had rarely been seen

---

[61] Interview with Rick Hauck by Mark Bowles, 23 August 2001.

[62] "Charter Planetary Panel Galileo and Ulysses Mission, Kennedy Space Center," Box 2142, Division STED Project Management Office, Folder Kennedy Correspondence, NASA GRC Records.

[63] Vernon Weyers, "Centaur Countdown," *Lewis News* (9 August 1985): 2, GRC Records.

since the glory days of the Apollo program. The theme from *Star Wars* accompanied the applause of more than 300 officials from the company, the Air Force, and NASA. Craig Thompson, a General Dynamics operations representative, said, "I almost had tears in my eyes when they rolled that thing out."[64] Alan Lovelace proclaimed it the "future of space probes in our generation."[65] The vehicle then underwent three weeks of extensive checking to make sure that all specifications and requirements were met. After passing all of the tests, SC-1 was flown to Cape Canaveral, where the crucial CISS was waiting for its mate. The CISS had been flown to the Cape two months earlier. It was placed on a converted Atlas-Centaur launch pad, where it received its final assembly and checks.

In September 1985, with eight months to go, Vernon Weyers, Deputy Manager of the Lewis Shuttle/Centaur Project Office, wrote, "All activities are in high gear in preparation for the dual planetary launches in May/June 1986."[66] There were some minor delays in assembling and mating the SC-1 with the CISS-1. There was a shortage of parts that were supplied by General Dynamics; this shortage resulted in about a one-month delay. However, Weyers was confident that this was not a problem because there was a two-month period built into the schedule for such common delays. The only trouble was that this luxury was not available for the next launch. The SC-2 and CISS-2 were right behind the first vehicle and did not have this schedule slack, so future delays could not be tolerated.

A month later, progress was still on track, and Larry Ross reported, "A lot of hard work remains, but all of the tasks required to launch the first Shuttle/Centaur seven months from today are doable."[67] General Dynamics sent the missing parts to Cape Canaveral, and the SC-1 and the CISS-1 were successfully mated together. A test rotation and separation of the Centaur was also completed successfully. Engineers hoped that this was the last time that this test would be performed before the real separation took place in orbit above Earth. SC-2 and CISS-2 were continuing their final inspections at General Dynamics in San Diego and would be transported across the United States in November. To reduce the difficulty in housing both the SC-1 and SC-2 at Kennedy Space Center (KSC) at the same time, the Air Force made their Shuttle Payload Integration Facility available between November and December. This gave NASA the ability to process both vehicles simultaneously. Other good news included the report of a 109-percent engine thrust level for the Galileo mission (100 percent was considered the optimum safety level, and the previous high was 104 percent). This thrust capability made it possible to launch a heavier Centaur that carried more propellant. With more energy, there were more avail-

---

[64] Michael L. Norris, quoted in "New Booster Rolled Out in San Diego," *The Los Angeles Times* (14 August 1985).

[65] Vernon Weyers, "Centaur Countdown," *Lewis News* (6 September 1985): 3, GRC Archives.

[66] Vernon Weyers, "Centaur Countdown," *Lewis News* (4 October 1985): 3, GRC Archives.

[67] Vernon Weyers, "Centaur Countdown," *Lewis News* (1 November 1985): 5, GRC Archives.

able launch days. Marshall engineers were actually displeased by this because they did not want to operate their engines at this risky thrust level. A higher thrust level pushed the engines past safe performance levels. The higher the accepted percentage level, the more powerful the thrust—and the higher the risk.

With six months to go, the SC-1 and the CISS-1 were still undergoing testing at the Space Center Eastern Test Range at Kennedy Space Center. Special attention was given to leak checks on the hydraulic system. A few minor leaks were detected, but all were tracked down and repaired. The next key test was the cryogenic cold flow test that was held in mid-November. Two successful tests were performed for both the liquid-oxygen and liquid-hydrogen systems. SC-1 and CISS-1 also passed the first two phases of the Design Certification Review. The final stage was the presentation of the certification to NASA Associate Administrator Jesse Moore.[68] While the first Shuttle/Centaur was being tested, its twin arrived in Florida. The CISS-2 was the first component to show up, and it waited for its mate. The SC-2 was still in California, where General Dynamics was performing propellant tank insulation and avionics system checks.

In January 1986, Centaur encountered two significant technical problems. During a test of the mounts that attached the Propellant Level Indicating System inside the liquid-oxygen tank, the mounts failed. Engineers quickly redesigned them, then fabricated and installed them on the SC-1. The new mounts passed all subsequent tests. The second main problem occurred during the liquid-oxygen and liquid-hydrogen cold flow tests. Specifically, the valves on the CISS-1, which controlled the fill, drain, dump, and engine feed functions, experienced erratic operations. Using a simulated system, engineers located this problem as well and developed a redesign. New tests were scheduled, and 15 January became a "do or die" hurdle for the program.[69] Both propellant tanks were completely filled and pressurized at 110 percent of flight pressure, and all of the major system procedures were tested. Vernon Weyers recounted, "Nobody close to the program dared to predict the success which was achieved . . . the entire avionics system performed flawlessly."[70] Robbins was equally pleased with the performance and concurred that a "very important milestone was completed."[71] Ross said, "The tanking was absolutely perfect."[72]

The astronauts were not as enthusiastic. In early January 1986, they became concerned about whether the abort of a Shuttle/Centaur mission could be safely orchestrated. They

[68] Vernon Weyers, "Centaur Countdown," *Lewis News* (29 November 1985): 4, GRC Archives.

[69] Vernon Weyers, "Centaur Countdown," *Lewis News* (10 January 1986): 3, GRC Archives.

[70] Vernon Weyers, "Centaur Countdown," *Lewis News* (7 February 1986): 3, GRC Archives.

[71] W. H. Robbins, "Shuttle/Centaur Weekly Status Report," 13–17 January 1986, Box 2143, Division STED Project Management Office, Folder Status Reports, NASA GRC Records.

[72] Interview with Larry Ross by Mark Bowles, 1 March 2000.

questioned whether there was adequate helium pressure to activate the dump valves if something went wrong. They worried that there were not adequate backup systems to ensure their safety, and they began to refer pejoratively to Centaur as "Death Star."[73] Astronaut Rick Hauck said, "This was a big issue for us. Very big." Hauck and John Young, the head of the Shuttle office, took up the safety issue with the Johnson Space Center Configuration Control Board. They were surprised when the Board ruled the system acceptable. Hauck recalled, "I went back to my crew that day and said, 'Things have changed at NASA. We are willing to take risks that we didn't take before and if any of you want to resign from this flight I will support you.'"[74] However, when given a chance to withdraw, with no questions asked, from the crew that was assigned to fly Shuttle/Centaur, none of the selected crew accepted the offer. Three weeks later, however, the *Challenger* disaster radically altered the perception of risk.

On 28 January, *Challenger* lifted off from Cape Canaveral; minutes later, it exploded over the Atlantic Ocean, thereby killing the crew. This was the worst disaster in NASA history up until that time, and the first time that any American astronaut was ever lost in flight. Immediately, the Shuttle/Centaur program was in jeopardy. Despite years of work, the rocket quickly began to fall out of favor with both Johnson Space Center and NASA Headquarters.

Fighting for Survival

The Centaur team at Lewis was devastated. Since Shuttle/Centaur was so close to the date of its first launch, many of the Lewis engineers working on the project were already at the Cape and witnessed the *Challenger* explosion with their own eyes. They immediately realized that Shuttle/Centaur, along with Galileo and Ulysses missions, was in jeopardy. When Red Robbins returned to Cleveland the next Monday morning, he went directly to the personnel office and tendered his resignation. Already close to retirement, Robbins knew that at best, Shuttle/Centaur would be years from launching, and he could not endure the wait.

Although the Shuttle/Centaur program was not immediately canceled, its future was very uncertain. On 20 February 1986, Jesse Moore sent out the order to postpone the Galileo and Ulysses launches.[75] The very earliest that the two missions could be rescheduled was thirteen months away. This was the time when Jupiter would be on the direct opposite side of the Sun from Earth. Even the Ulysses mission was dependent upon the location of Jupiter

---

[73] Interview with Rick Hauck by Mark Bowles, 23 August 2001.

[74] Ibid.

[75] Jesse Moore, "Galileo and Ulysses Launch delay," 20 February 1986, Box 2143, Division STED Project Management Office, Folder Status Reports, NASA GRC Records.

because it was going to receive a gravity-assist boost from the planet to catapult its way to the Sun. In the meantime, the Galileo probe was moved to the Vertical Processing Facility at KSC, where it was mated with the Centaur. Engineers continued to perform compatibility and separation tests in the hopes that the missions might still get a green light.[76]

Acting NASA Administrator William R. Graham said that his decision to postpone did not mean that the next Space Shuttle launch would be delayed until after the Galileo and Ulysses launch opportunity the next May. The Shuttle might be back in operation earlier than thirteen months, but he said that two key factors forced his hand to delay these missions: 1) key personnel required to ensure the safe and successful launch of either Galileo or Ulysses were preoccupied with the timely analysis of causes of the 51-L accident and 2) the consequences of the accident had significantly eroded the schedule margins for launch-site processing required prior to the first flight of the Shuttle/Centaur upper stage.[77]

Despite guarded optimism, even if an additional thirteen months of testing could be performed, the hopes for an eventual Shuttle/Centaur launch were beginning to fade. One reason for concern was the issue of weight and the ongoing disagreement between Lewis and Johnson. The Lewis research team had three key weight factors to consider: the weight of Centaur, the weight of its payload (Galileo), and the weight of the Centaur propellant. When Johnson committed to a 65,000-pound lift capability for the Shuttle, Lewis knew that they were slightly over the weight allowance. However, the actual lift capability of the Shuttle was never close to that commitment. To compensate, Centaur was able to decrease the amount of propellant carried in its tanks. This tradeoff meant that the launch window shortened because Centaur would have insufficient fuel to hit the planetary target on those days that required a higher spacecraft velocity at separation.

Early in the integration process, Lewis engineers struggled with the weight problem because Johnson engineers kept decreasing their allowance for the Centaur weight. Approximately two weeks before the *Challenger* launch, Lewis trajectory engineer Joe Nieberding made a presentation to Jesse Moore and to Lewis, KSC, JSC, and MSFC top management sounding the alarm that the number of viable launch days had decreased to fewer than six for Galileo, far too few to ensure a reasonable probability of launch. In the presentation, he demonstrated that using 109-percent Shuttle engine thrust (instead of the previously high level of 104 percent) was the only potentially viable way of restoring a reasonable number of launch days. In fact, Jesse Moore had approved the higher thrust level on the spot, over the strenuous objections of JSC and MSFC. So when *Challenger* exploded,

[76] "Galileo, Centaur Milestone Tests are Underway," *Lewis News* (16 May 1986): 2, NASA GRC Archives.

[77] James F. Kukowski, "NASA Postpones Galileo, Ulysses, Astro-1 Launches," NASA Headquarters Press Release, 10 February 1986, GRC Public Affairs Archives; William R. Graham, "Galileo and Ulysses Missions Postponed for Year," *Lewis News* (21 February 1986): 6, NASA GRC Archives.

Lewis engineers knew that any redundancies added to make the Shuttle safer would also make it heavier. It would be able to carry even less than before. Although in no way related to the failure, the fact that *Challenger* exploded only a few seconds after its engines reached 104 percent of its nominal thrust contributed to the perception that pushing the Shuttle engines to 109 percent on the Galileo mission would be unsafe.[78] With the Shuttle getting heavier and able to lift less weight when it returned to flight, and with JSC and MSFC now vehemently opposed to pushing the Shuttle engines above a 104-percent thrust level, Lewis engineers realized that the Shuttle would never be able to lift Centaur and Galileo.[79]

Another issue was safety as it related to Centaur's fuel. Following the Shuttle disaster, Lewis held meetings on the Centaur safety program in May 1986. The attendees included representatives from General Dynamics, JSC, KSC, the Air Force, TRW, Boeing, Lockheed, Martin Marietta, and Analex, an aerospace industry service provider located in Brookpark, Ohio. The goal of the meeting was to prove that despite the risk presented by liquid hydrogen, the Centaur rocket was safe. Engineers described the entire safety organization for the Centaur at Lewis, reviewed recent hazard reports, and presented a revised abort/contingency plan.[80] But the effort was not enough to convince opponents.

Even more serious, from the point of view of engineers at Lewis, was the realization that increased redundancy in Centaur and Shuttle systems meant that the Shuttle would no longer be able to lift Centaur. They knew that the Shuttle engines would not be allowed to push beyond the 100-percent thrust level, significantly less than the 109-percent level needed.

Another problem that contributed to the growing pessimism over the future of Centaur concerned General Dynamics. Despite a long and successful relationship, Lewis management was coming to believe that General Dynamics no longer possessed the requisite skills to build and staff Centaur effectively. The astronauts were especially critical.[81] When astronaut David Walker went out to the plant, he was surprised at the conditions he found. Compared to the Rockwell facility, where the Shuttle was meticulously assembled in a clean, white-glove environment, there seemed to be less attention to manufacturing a quality product at General Dynamics. Larry Ross recalled one incident when there was a lapse in common sense on the part of workers moving a Centaur from the factory to a test site. "They took a Centaur G and ran it into a bridge going up to Sycamore Canyon," he said.

---

[78] Murray, *Journey Into Space,* 234.

[79] Interview with Joe Nieberding by Virginia Dawson and Mark Bowles, 15 April 1999.

[80] William E. Klein, "Minutes for Shuttle/Centaur G Safety Certification Process Briefing," 2 June 1986, Box 2142, Division STED Project Management Office, Folder Technical Paper Approvals by PM, NASA GRC Records.

[81] Interview with Rick Hauck by Mark Bowles, 23 August 2001.

"It was dumb. They hadn't measured the height." He felt that some of the fault lay in Lewis's management. "The lesson was that we should have fired some people at General Dynamics. That was a mistake."[82] Vernon Weyers also was aware of the problems at General Dynamics. He visited the plant once or twice a month over the four-year Shuttle/Centaur program. He suspected that drugs were a problem. He recalled, "I personally was out in the shop by myself at General Dynamics one day where they were working on a big piece of structure. I looked through the access hole, and an engineer was sound asleep. It was very frustrating at times, because on the technical side, the production side, they did not seem to get their act together . . . . They just did not perform well technically."[83]

On 22 May 1986, Headquarters began to address the problems surrounding Shuttle/Centaur and consider what future action they should take. Rick Hauck gave an influential briefing citing risks to the Shuttle crew as the reason why the program should be canceled. He concluded that "Shuttle/Centaur, even after satisfactory accomplishment of currently identified safety-related fixes, poses additional risk to crew and Shuttle safety. Attempts to integrate an unmanned upper stage in the Shuttle have resulted in compromises, which caused undue risk to orbiter and crew. The ability to reduce these risks to an acceptable level is questionable."[84] An independent study of safety concerns, conducted by the House Committee on Appropriations and chaired by Edward P. Boland and Congressman William Green, concluded that it was in the best interest of NASA to terminate the Shuttle/Centaur program.

On 19 June, NASA Administrator Dr. James C. Fletcher gave the order to terminate the program.[85] After the country had spent nearly one billion dollars to transform Centaur for the Shuttle, Lewis Research Center received this heartbreaking order: "You are directed to terminate the Shuttle/Centaur upper stage program."[86] Fletcher stated, "Although the Shuttle/Centaur decision was very difficult to make, it is the proper thing to do and this is the time to do it."[87] Astronaut and future NASA Administrator Richard H. Truly was directed to examine other alternatives for the Shuttle/Centaur planetary and scientific payloads.

The result was catastrophic for the space program. Suddenly, the Galileo probe again lost its ride into space. Not only was it forced to wait until a solid rocket upper stage could be built, but also, the lack of power meant that the entire mission would take much longer.

---

[82] Interview with Larry Ross by Mark Bowles, 1 March 2000.

[83] Interview with Vernon Weyers by Mark Bowles, 8 April 2000.

[84] Interview with Rick Hauck by Mark Bowles, 23 August 2001.

[85] John M. Logsdon, "Return to Flight: Richard H. Truly and the Recovery from the *Challenger* Accident," in *From Engineering Science to Space Science*, 360.

[86] "Shuttle/Centaur Termination Status" NASA Headquarters, 4 September 1986, Box 2144, NASA GRC Records.

[87] Sarah G. Keegan, "NASA Terminates Development of Shuttle/Centaur Upper Stage," NASA Headquarters Press Release, 19 June 1986, NASA GRC Public Affairs Archives.

With a Centaur launch, Galileo would have arrived at Jupiter in two years. When Galileo finally did launch in October 1989 with the less powerful Boeing-built Inertial Upper Stage, it took six years to arrive in the Jovian atmosphere.[88]

The loss of Centaur also dramatically changed the navigational course that Galileo took. With Centaur, the Galileo flightpath was to have been direct. With the solid IUS, Galileo was required to take a much more circuitous route with the help of gravitational assists. Galileo actually flew by Venus once and Earth twice, each time using the planet's gravity to effectively "steal" some energy and direct it to the velocity of the craft. Although this reduced the duration of the flight, the six years to arrive in the Jovian atmosphere still took three times as long as what Centaur was capable of delivering.

Not only did the Centaur cancellation result in nearly a decade's delay of Galileo for data return, but it also significantly jeopardized mission success. In April 1991, Galileo attempted to open the large high-gain antenna that was the primary mechanism by which the craft transmitted Jupiter data to Earth.[89] Soon after the attempt, telemetry revealed to the JPL engineers that something had gone wrong. The motors only partially opened the antenna, and then they stalled. Over a period of weeks, more than a hundred experts at JPL analyzed the problem and concluded that the antenna failed because of excessive friction between antenna pins and sockets. Ironically, JPL engineers blamed this problem on the cancellation of Centaur. In 1985, Galileo and the antenna were shipped by truck from JPL to Kennedy. After the *Challenger* explosion, Galileo was shipped back across the country in 1986, and then back again in 1989, when it was finally launched. The vibration that the antenna experienced during the cross-country trips by truck loosened the pin lubricant and caused the main antenna to fail.

Although some important scientific observations were lost, the Galileo mission remained a success when engineers reconverted the low-gain antenna.[90] In October 1991, Galileo performed a flyby operation of the 8-mile-diameter asteroid Gaspra. When Galileo passed by at a distance of 1,000 miles, it was the first time a spacecraft had provided direct observations of an asteroid.[91]

Like Galileo, the Ulysses mission also suffered from the cancellation of the Shuttle/Centaur program. D. Eaton was the Ulysses project manager at the European Space Agency. Upon hearing news of the suspension of the Centaur program, he and his team reacted with "deep

[88] Barbara Selby, Leon Perry, and Terry Eddleman, "Upper Stage Selected for Planetary Missions," NASA Headquarters Press Release, 26 November 1986, NASA GRC Public Affairs Archives.

[89] Paula Cleggett-Haleim, "Galileo Antenna Deployment Studied by NASA," NASA Headquarters Press Release, 19 April 1991, GRC Public Affairs Archives.

[90] "New Telecommunications Strategy Aimed at Maximizing Return from Galileo's Low-Gain Antenna," *http://cass.jsc.nasa.gov/publications/newsletters/lpib/lpib76/gal76.html*.

[91] Paula Cleggett-Haleim and Jim Wilson, "Galileo To Set Course for Encounter with Asteroid Gaspra," NASA Headquarters Press Release, 28 June 1991, NASA GRC Public Affairs Archives.

disappointment." Despite the wasted money and the delay to the Ulysses program, Eaton said that "the overriding thought is the bitter blow this must be to all at Lewis and General Dynamics who worked intensely long hours to achieve an impossible schedule only to have the goal posts taken away and the game abandoned at the last minute."[92] Like Galileo, Ulysses also eventually made it into space. In May 1991, the spacecraft took its long journey to Jupiter for an eventual slingshot to the Sun. While at the planet, it performed physics investigations such as taking measurements of the Jovian magnetosphere.[93]

To engineers at Lewis Research Center and JPL, however, the abandonment of the program was not in the national interest. Nieberding said, "To this day, the country is hurting for not having a liquid-hydrogen upper stage coupled with the Shuttle or with an equivalent expendable launch vehicle capability."[94] Tom Shaw at JPL recalled, "June 19, 1986, was a dark day as far as I was concerned. I felt that we suffered a truly significant setback in national capability, and specifically my own narrow purview of planetary exploration, where we need high-energy vehicles to get there with any kind of significant payload . . . . You learn how to use a much, much smaller spacecraft."[95]

Beyond their disappointment over the cancellation of Shuttle/Centaur, Lewis Research Center engineers pointed out that many missions carried by the Shuttle in the 1980s, including the commercial satellite payload carried by the ill-fated *Challenger*, should have been assigned to an expendable launch vehicle such as Atlas-Centaur. Red Robbins observed, "I think they sold their soul to the devil to fund the Shuttle . . . the expendable launch vehicle was a good animal and not nearly as expensive as the Shuttle program." He said that while the expensive Saturn V could deliver approximately 250,000 pounds of payload to low-Earth orbit, the Shuttle, which eventually cost as much, could deliver the same weight into space. The problem was that 200,000 pounds of that weight was the Shuttle itself. Thus, with the Shuttle payload of only 50,000 pounds, "the cost per pound to fly into low-Earth orbit with the Shuttle is up by a factor of 5. You don't even need a rocket scientist to tell you that."[96]

---

[92] D. Eaton to Larry Ross, 11 July 1986, Box 2143, Division STED Project Management Office, Folder Galileo Correspondence, NASA GRC Records.

[93] Paula Cleggett-Haleim and Robert MacMillin, "Ulysses To Begin Jupiter Physics Investigations," NASA Headquarters Press Release, 28 May 1991, NASA GRC Public Affairs Archives.

[94] Interview with Joe Nieberding by Virginia Dawson and Mark Bowles, 15 April 1999.

[95] Interview with Tom Shaw by Virginia Dawson, 9 November 1999.

[96] Interview with Red Robbins by Mark Bowles, 21 March 2000.

Terminating the Program

Ends can come for many different reasons. Technologies can evolve to the point where older designs are obsolete. Mission objectives can change, making previous areas of exploration unnecessary. Costs can escalate to the point where funding no longer allows the continuation of a project. Managerial and technical errors can jeopardize a mission. Although these are the most common reasons for premature endings to NASA projects, Shuttle/Centaur did not encounter any of these fates. The technology was still cutting-edge; planetary missions were still needed; the funding remained intact; and no uncorrectable errors were made in either the design or the management of the program. Shuttle/Centaur met with one of the most devastating and uncontrollable problems—fear. After the *Challenger* explosion, a liquid-hydrogen rocket inside the bay doors of the Shuttle just looked too dangerous and too risky for NASA to continue with the program.

In June 1986, NASA sent out termination letters to their own Centers and to contractors like General Dynamics, Honeywell, Analex, Teledyne, and Pratt & Whitney. Their order was, "You are hereby directed to take those necessary steps for an orderly shutdown and cessation of all work."[97] Each of these organizations had to slowly bring Centaur work to a halt, resulting in over 200 stop-work orders.[98] With great remorse, Andy Stofan wrote, "Lewis is proceeding with the orderly shutdown of the terminated Shuttle/Centaur program."[99] Throughout July, "closeout meetings" were held with all the contractors. Most of the work of the contractors was completed by 30 September 1986; all work was to be finished by the end of the year. Each contractor was expected to set a completion date for itself that would maximize the potential use of its work for the government. Because Centaur might be used again as an expendable launch vehicle, all contractors were also instructed to assess the impact that the end of the Shuttle/Centaur program would have on future capabilities.[100]

The remaining resources from the Shuttle/Centaur program included documentation, technology, and people. The documentation was the easiest resource to relocate. Fortunately, in February 1986, NASA created the Centaur Engineering Data Center where all engineering paperwork associated with the project was sent. Besides increasing the speed

---

[97] Termination letter 23 June 1986, "Shuttle/Centaur Termination Status" NASA Headquarters, Box 2144, NASA GRC Records.

[98] R. E. Jumont to W. H. Robbins, 16 July 1985, Box 2142, Division STED Project Management Office, Folder General Dynamics Correspondence, NASA GRC Records.

[99] Andrew J. Stofan memo to R. Truly, 8 July 1986, "Shuttle/Centaur Termination Status" NASA Headquarters, Box 2144, NASA GRC Records.

[100] Vernon J. Weyers to distribution, 1 July 1986, Box 2142, Division STED Project Management Office, Folder General Correspondence, NASA GRC Records.

in which information could be accessed, it was also designed to relieve engineers of filing chores and promised to free up work space.[101] This information included official files stored in personnel offices, the mailroom, contractor sites, and storage facilities.[102] The Data Center received key documents, and the information was entered into a database and then sent to a vault. Many materials were then destroyed as engineers cleaned out their offices. Saved materials included minutes of all reviews of the project, significant correspondence, and all engineering drawings. Between October and November 1986, all the major contractors and NASA Centers began the process of turning over their documentation.

By the middle of 1987, most of the contractors had issued final reports on various aspects of the Shuttle/Centaur development. In May 1987, Pratt & Whitney released a final report on the "RL10 Ignition Limits Test for Shuttle Centaur." Kennedy Space Center and Lewis Research Center summarized testing done to determine instability in the CISS and the Pneumatic Activated Valve Control System. United Technologies issued a final report called "Shuttle Centaur Engine Cooldown Evaluation." These three reports were representative of the type of work done to resolve technological questions that still surrounded the project.[103]

Preservation of the technologies associated with the Shuttle/Centaur project was a more difficult problem. Suggestions on how to use the technology were made by engineers who could not accept the possibility that their years of work had been for naught. For example, Floyd Smith, the chief of the structures and facilities branch at Lewis, believed that the government should review all potential hardware uses and then store the equipment in a suitable location. He hoped that it might still be possible to use one of the new Centaur vehicles for a later NASA mission. Even if NASA decided never to use the new rocket configuration, Smith wanted other possible spinoffs from their work. He thought that one application might support other payload mounting structures in the Shuttle.[104]

The Air Force also had to decide what to do with its Shuttle/Centaur hardware. Because the Air Force was a full partner in the development costs, it was entitled to any of the residual hardware that might be used for other craft, like Titan-Centaur.[105] In June 1986, their official termination orders stated, "Through appropriate coordination with NASA

[101] James O. Rogers, Jr., to distribution, 20 February 1986, Box 2142, Division STED Project Management Office, Folder General Correspondence, NASA GRC Records.

[102] Thomas S. Banus to distribution, 11 June 1986, NASA Archives, Box 2142, Division STED Project Management Office, Folder General Correspondence; Thomas S. Banus to distribution, 2 July 1986, Box 2142, Division STED Project Management Office, Folder General Correspondence, NASA GRC Records.

[103] Final reports found in Box S/C Records, consignor Alan Willoughby, NASA GRC Records.

[104] R. C. Edwards for Floyd Z. Smith to Acting Manager, Shuttle/Centaur Project Office, 21 July 1986, Box 2142, Division STED Project Management Office, Folder General Correspondence, NASA GRC Records.

[105] C. Thomas Newman to Administrator, 29 September 1986, Box 2142, Division STED Project Management Office, Folder Air Force Correspondence, NASA GRC Records.

Headquarters and in view of the needs of the DOD, you are to implement the shutdown in a manner which is most advantageous to the U.S. Government."[106] To take greatest advantage of the work already accomplished, NASA asked for additional funds to enable its contractors to take their hardware development to a suitable stopping point. Thus, the completed technology would have a greater chance of being used on another spacecraft.

In the process of terminating the program, Lewis Research Center received a number of accolades from the Air Force. Larry Ross received a letter of appreciation from Colonel David Raspet at the Department of Defense in recognition of the "exceptional performance" of his laboratory. Raspet said that because of the efforts of Lewis, despite the cancellation of the program, "We were on course with Centaur G and succeeded in laying foundations for Titan/Centaur."[107] In December 1986, Brigadier General Nathan J. Lindsay said that the government had decided that it was in the government's best interest to buy back all of the flight hardware from NASA. He believed that the "termination has been difficult for all of us,"[108] but he offered his thanks to Lewis Research Center for helping to make the process as smooth as possible. Ross appreciated the accolades given to him and his organization by the Air Force and wrote, "Dispositioning these assets is a painful process, but I find consolation in the potential for their use on other flight vehicles."[109]

Centaur G-prime lived on when it was mated, with some modifications, to the Air Force Titan IV booster, first launched in 1994. In 1992, NASA chose Titan IVB/Centaur G-prime as the launch vehicle for the Cassini mission to Saturn launched in 1997. This $3.4-billion mission, a project managed jointly by NASA, the European Space Agency (ESA), and the Italian Space Agency (Agenzia Spaziale Italiana, ASI), carried a huge nuclear-powered probe programmed to land on the surface of Titan, the largest of moon of Saturn. The return of Titan-Centaur, the last and most expensive of NASA's large spacecraft, underlined the appropriateness of expendable launch vehicles for ambitious space science missions that required the power and precision of Centaur.

A Changing Tolerance for Risk

President Nixon had closed his 1972 announcement of the Shuttle program with a prophetic quotation from Oliver Wendell Holmes: "We must sail sometimes with the wind

---

[106] Vernon J. Weyers, Acting Manager Shuttle/Centaur Project Office, to Headquarters Space Division, 26 June 1986, Box 2142, Division STED Project Management Office, Folder Air Force Correspondence, NASA GRC Records.

[107] David Raspet to L. J. Ross, 4 November 1986, Box 2142, Division STED Project Management Office, Folder Air Force Correspondence, NASA GRC Records.

[108] Nathan J. Lindsay to NASA Lewis Research Center, 9 December 1986, Box 2142, Division STED Project Management Office, Folder Air Force Correspondence, NASA GRC Records.

[109] Larry Ross to Major General Ralph H. Jacobson and Brigadier General Nathan J. Lindsay, 30 October 1986, Box 2142, Division STED Project Management Office, Folder Air Force Correspondence, NASA GRC Records.

and sometimes against it, but we must sail and not drift nor lie at anchor."[110] Few other statements could have so accurately captured Centaur's relationship with the Space Shuttle.
Implicit within Holmes's words was a notion of risk. Sailing is a dangerous endeavor. When
the wind blows at our back, the chance of safely completing the voyage is increased.
However, when we set sail with the gusts directly in our face, churning the waters against
the ship's bulkhead, the conditions for ultimate success are reduced. Holmes said that
inherent in the nature of humanity is the desire to raise anchors and explore the unknown.
We endure the dangers. We accept the risks of what we do because we believe that the
rewards ultimately justify the sacrifice.

Holmes's eloquent argument for the adventurous spirit suggests that risk is not a
constant factor, equally endured by everyone. The tolerance for risk changes like the wind,
deviating from one person, organization, or era to the next. What one person might consider
an acceptable risk assumed to achieve some larger goal, a person with a more conservative
outlook might reject for one that yields less reward, but also less potential for failure.

Safety concerns are now given as the official reason for termination of the
Shuttle/Centaur program. One author reported in 1999 that this Shuttle configuration was
"eventually scuttled because of safety concerns stemming from the Centaur's common-bulk-
head tank design and from the difficulties of dumping propellants in an abort situation."[111]
However, blaming the entire failure upon technical issues related to safety obscures the engineering triumphs that made Shuttle/Centaur flight-ready. These safety concerns were
nothing new and did not emerge suddenly after the tragic *Challenger* loss. Instead, they were
evident from the very start of the program and were embedded in the Centaur design. Many
engineers considered that the tank design and propellant dumping concerns were part of the
acceptable risks of flight. They worked many years to reduce the risks even further to meet
NASA standards. In this they were successful. Before the *Challenger* tragedy, General
Dynamics confidently proclaimed, "The proven Centaur System can now provide maximum
benefits to Shuttle users."[112] Vernon Weyers concluded, despite the technical challenges and
problems encountered, "At the time of the *Challenger* accident . . . there was little doubt that
the Centaur would be ready on time to support the 1986 dual planetary launches."[113] Marty

---

[110] White House Press Secretary, "The White House: Statement by the President," 5 January 1972, Richard M. Nixon Presidential Files, NASA Historical Reference Collection, NASA Headquarters, Washington, DC, as found in "Nixon Approves the Space Shuttle," in *NASA: A History of the U.S. Civil Space Program*, ed. Roger D. Launius (Malabar, Florida: Krieger Publishing Company, 1994), 235.

[111] Ivan Bekey, "Exploring Future Space Transportation Possibilities," in *Exploring the Unknown, Volume IV*, ed. John M. Logsdon (NASA SP-4407, 1999), 509.

[112] General Dynamics, Box 2146, Division STED Project Management Office, Folder S/C Marketing Vugraphs from GDC, NASA GRC Records.

[113] Vernon Weyers, "Development of the Shuttle/Centaur Upper Stage," paper delivered at the IAF Congress, Innsbruck, Austria, October 1986, Box 2143, Division STED Project Management Office, Folder IAF Paper, NASA GRC Records.

Winkler of General Dynamics concurred with this assessment: "I'm convinced to this day we would have made the launch window in May of '86, but it was a sprint to the finish. It was like the racehorse that overtakes you at the end. Had not the Shuttle accident occurred, we would have been ready for launch."[114] So, if technically the Centaur would have been ready, something beyond safety was at play in the decision to terminate.

The first reason was a monetary issue. NASA management knew that the effort required to return the Space Shuttle to flight-ready status would be enormously expensive. As a result, there would not be enough left in the budget to work out any remaining technical issues with Shuttle/Centaur. Marty Winkler said, "it wasn't canceled because of technical problems, although there were technical problems, because everyone agreed that the technical problems were solvable. It was canceled because NASA had concluded internally . . . that they didn't have enough manpower, energy, and money to fix both the Shuttle's return to flight and the Centaur. That's a little-known fact, but it's absolutely true."[115] Also, a repaired Shuttle would never have had the lift capability to launch a "safer" and heavier Centaur.

Another component was even more central to the cancellation of the program—a changing tolerance of risk. Liquid hydrogen was such an alluring fuel because of the capabilities that it gave to explore space. The tradeoff for greater power and control was heightened safety concerns. Liquid hydrogen was dangerous, and there was never any question that it was more volatile than solid propellants. However, during the 1970s and early 1980s, these risks were considered acceptable. Ironically, the *Challenger* explosion was caused by the failure of one of the solid-propellant booster rockets—previously considered very reliable. The tragedy brought with it a new era of conservatism and a diminished acceptance of risk. As Larry Ross argued, "After *Challenger* the equation changed."[116] The design specifications that had been acceptable before *Challenger* were based upon theoretical and technical judgments. In the wake of the explosion, new "emotional" judgments were added to the mix. Because of these strong emotional factors, Ross concluded, "There was no crew that would fly it because they were staggered by *Challenger*. It was not rational; it didn't have to be rational. It was human life."[117]

The astronauts agreed that the level of acceptable risk changed, and that NASA had been willing to assume too many risks in the pre-*Challenger* days. Rick Hauck said that the changing levels of risk tolerance were a part of NASA's history. He said, "It is all part of the ebb and flow of the desire to succeed and the desire to meet schedules. Probably there was

---

[114] Interview with Martin Winkler by Virginia Dawson, 21 March 2001.

[115] Ibid.

[116] Interview with Larry Ross by Mark Bowles, 1 March 2000.

[117] Ibid.

a certain amount of confidence born of never having an in-flight failure before."[118] There was also a risk tolerance difference in 1986 between Johnson Space Center and Lewis Research Center that he called "cultural." He said that the Lewis engineers were "highly motivated professionals who believed they were exercising as much caution as needed to be exercised." But generally these engineers believed Centaur could fly in the Shuttle. As a whole, Lewis always believed that "the human spaceflight guys at Johnson were a bit too cautious." Hauck concluded, "There was a natural cultural difference between the groups. The issues about risk are tied up in the cultural issues."[119] While Hauck was on the side that supported grounding Shuttle/Centaur forever, he conceded that there was an irresolvable difference between the two sides. Eventually, the less risky solution won out.

Richard Kohrs was right in the middle of the debate between the engineers and the astronauts. Kohrs was the manager of the Technical Integration Office for the Space Shuttle from 1980 to 1985 at Johnson Space Center. He was also the main point of contact between Lewis Research Center and Johnson, so he understood both cultures. He conceded that Shuttle/Centaur seemed like a good idea before they got to the "now let's go make it work" phase. But when they began to go through all of the battles concerning weight issues, single point failures, interfaces, redundancy, and so on, he and his colleagues at Johnson came to realize that the plan was less desirable. Although initially the risks seemed worth taking, the risk tolerance changed as the program progressed. Kohrs recalled, "I think a lot of these risks were accepted before . . . the [technical] guys had to go make it happen. Then of course you got into the flight crew concern of this whole thing." Johnson protected the astronauts, and because the astronaut crews united against the Shuttle/Centaur plan, Kohrs and his colleagues also withdrew support. Although Kohrs ultimately adopted the more conservative attitude of the astronauts, he still had faith that from a technical perspective, "I think it [Shuttle/Centaur] would have flown."[120]

Those engineers who devoted years of their lives to readying Centaur for the Shuttle were part of a culture with a tolerance for risk. They knew their rocket and they believed the mission would succeed. To the engineers involved in Shuttle/Centaur, the rocket represented hope for a new era of space exploration. Their enthusiasm did not blind them to safety problems, but they believed that they had reduced the risk to manageable levels for mission success and pilot safety. These engineers also correctly pointed out that the human risk factor should never have entered the equation in the first place. Joe Nieberding concluded, "These missions should have flown on an expendable rocket. They needed no

---

[118] Interview with Rick Hauck by Mark Bowles, 23 August 2001.

[119] Ibid.

[120] Interview with Richard Kohrs by Mark Bowles, 13 July 2001.

human crew. Again, the decision to fly everything on STS was a disaster."[121] Historian William Burrows agreed with this assessment of the tragic loss of life in the deployment of a satellite that could easily have been launched with an expendable rocket.[122]

In 1990, a NASA advisory committee produced a report on the future of the United States space program and discussed the significance of risk for the Agency. The committee concluded that risk has always been a central feature of all of the greatest human adventures. When Magellan first circumnavigated Earth in 1519, he started his voyage with five ships and a crew of 280. After the three-year voyage, only one ship and thirty-four crewmen returned. Early test pilots in the 1950s faced a similar risk factor while pushing the limits of supersonic flight. The committee wrote, "Risk and sacrifice are seen to be constant features of the American experience. There is a national heritage of risk taking handed down from early explorers . . . it is this element of our national character that is the wellspring of the U.S. space program."[123]

Despite this bold statement of our adventurous heritage, NASA worried that the "spark of adventure is flickering. As a nation, we are becoming risk averse."[124] NASA knew this from experience. Risk aversion had kept them from launching Shuttle/Centaur. Richard H. Truly, Associate Administrator for space flight and charged with the responsibility to return the Space Shuttle safely to flight, wrote, "I know that the business of space flight can never be made to be totally risk-free, but this conservative return to operations will continue . . . ."[125] But risk is a subjective, ever-present force that waxes and wanes in the minds of all who step into a NASA program. NASA and America would recover from the *Challenger* tragedy. They would never forget their fallen colleagues, but they would move on and again accept the risk inherent in the exploration of space. Left behind was Shuttle/Centaur. It would forever remain an untested dream, eclipsed by a tragedy not of its own making. Marty Winkler expressed the unease that settled over the Centaur program in the wake of the Shuttle/Centaur cancellation. He said, "We were absolutely consumed through the early to mid-80s with Shuttle/Centaur. It sucked up all our energy. And when the program was canceled, we said, 'What do we do now?'"[126]

---

[121] Joe Nieberding correspondence with authors, 7 November 2001.

[122] William E. Burrows, *This New Ocean: The Story of the First Space Age* (New York: Random House, 1998), 556.

[123] *Report of the Advisory Committee on the Future of the U.S. Space Program* (NASA publication, December 1990), 16–17.

[124] Ibid.

[125] Richard H. Truly to NASA distribution, "Strategy for Safely Returning the Space Shuttle to Flight Status," 24 March 1986, in *Exploring the Unknown, Volume IV*, ed. John M. Logsdon (NASA SP 4407, 1999), 378.

[126] Interview with Marty Winkler by Virginia Dawson, 21 March 2001.

Chapter 8

Like a Phoenix

*"The sale of one commercial launch by a United States company is equiv-
alent to the import of ten thousand Toyotas."*

–Congressman Bill Nelson

Like a phoenix, Centaur was reborn in the 1980s and would remain a dependable and important upper stage launch vehicle into the new millennium. The source of this rebirth came from the commercialization of expendable launch vehicles. Just as it had survived the end of the Surveyor program in the 1960s and at the cancellation of the expendable launch vehicle program at the end of the 1970s, Atlas-Centaur emerged once again after the loss of *Challenger*. The impetus came this time from the communications satellite industry and the military.

In hindsight, it appears that NASA's decision in the 1970s to phase out all expendable launch vehicles in favor of the Shuttle was seriously flawed. This decision made it easier for Arianespace, a corporation backed by a consortium of European governments, to break the monopoly over space transportation that NASA had enjoyed since the 1960s. In the early 1980s, Arianespace became NASA's competitor for commercial satellite business, and other countries like China, Russia, and Japan followed somewhat later. The Europeans correctly predicted that this market would grow in the 1980s and 1990s and that the Shuttle would not be able to keep up with the demands of this market.[1] Ariane had a competitive edge. Because the new Arianespace launch facilities in Kourou, French Guyana, were closer to the equator, Ariane required a shorter trajectory to reach orbit than rockets launched from Cape Canaveral. Moreover, unlike launch facilities at Cape Canaveral, which had been allowed to deteriorate in anticipation of closing down expendable launch vehicle production, they were new and designed to serve the customer.

The reusable Shuttle was originally pitched to Congress as more economical than expendable launch vehicles. By flying the Shuttle often, it was thought that the cost per

---

[1] See Andrew J. Butrica, ed., *Beyond the Ionosphere: Fifty Years of Satellite Communication* (NASA SP-4217, 1997), especially Lorenza Sebesta, "U.S.-European Relations and the Decision to Build Ariane, the European Launch Vehicle," 137–156. See also Bruce D. Berkowitz, "Energizing the Space Launch Industry," Issues in *Science and Technology* 6 (Winter 1989–1990): 77–83.

launch would be significantly lower than that of an expendable launch vehicle that is used only once. The Shuttle, however, proved to be vastly more expensive than predicted and capable of far fewer flights per year. Moreover, a two-stage expendable rocket like Atlas-Centaur (or Ariane) is vastly more efficient than the Shuttle because it can carry a satellite directly to a geosynchronous transfer orbit. After separation, the satellite motor pushes it a relatively short distance to the desired geosynchronous orbit, where it moves at the same speed as the rotating Earth. By contrast, because the Shuttle is limited to low-Earth orbit (about 200 miles from Earth), a satellite launched from the Shuttle bay needs to use more of its own fuel to reach orbit, thus shortening its life as a communications relay.

Despite these significant drawbacks, NASA had pressed forward with its plans to make the Shuttle the country's only space transportation system. All three expendable launch vehicle manufacturers were in the process of closing down production. Termination of the Atlas-Centaur program was forestalled in 1980 when Intelsat, concerned over delays in Shuttle development, insisted on being able to use Atlas-Centaur for at least four more launches.[2] Advocates of Atlas-Centaur urged NASA to articulate a formal policy stating that it was in the national interest to allow expendable rockets to coexist with the Space Shuttle.[3] In the early 1980s, the Reagan administration took the first tentative steps toward the commercialization of expendable launch vehicles by setting up an Office of Commercial Space Transportation. However, expendable launch vehicle manufacturers found it impossible to compete against the highly subsidized Shuttle and the European Ariane rocket until the loss of *Challenger* precipitated a change in national space policy.

Global Communications Satellites: Intelsat, Comstar, FLTSATCOM

Although science fiction writer and visionary Arthur C. Clarke had imagined an important role for communications satellites as early as 1945, he had tempered this vision by calling attention to a technical hurdle that many thought might scuttle the dream of an artificial satellite communications system—the lack of a rocket powerful enough to launch a satellite into geosynchronous orbit.[4] Despite his grand hopes for satellite communications, he thought it would be many decades before the world would have a rocket with sufficient power to break the bonds of gravity. Twelve years later, the Soviet Union launched Sputnik, pushing the United States to develop a new satellite communications industry that would dominate into the next century.

---

[2] "Report of the Atlas/Centaur Review Board," February 1981, NASA Glenn Records.

[3] Ibid.

[4] Arthur C. Clarke, "Extra-Terrestrial Relays: Can Rocket Stations Give World-Wide Radio Coverage?" *Wireless World* (October 1945): 305.

The commercial satellite industry came into being when Congress passed the Communications Satellite Act in 1962 and created the quasi-commercial American Communications Satellite Corporation, or Comsat.[5] Several large American telecommunications companies such as AT&T equally owned Comsat, the corporate entity that managed the emerging global satellite system for the International Telecommunications Satellite Organization (Intelsat), organized in 1964. The system was designed to carry voice, telegraph, data, facsimile, and television transmissions. Although Intelsat was an international organization, it was heavily weighted to favor American interests.

Date	Mission	Vehicle	Payload Weight to Synchronous Orbit*	Result
25 Jan. 1971	Intelsat IV	AC-25	1,600 lbs	Success
19 Dec. 1971	Intelsat IV	AC-26	1,600 lbs	Success
22 Jan. 1972	Intelsat IV	AC-28	1,600 lbs	Success
13 June 1972	Intelsat IV	AC-29	1,600 lbs	Success
23 Aug. 1973	Intelsat IV	AC-31	1,600 lbs	Success
21 Nov. 1974	Intelsat IV	AC-32	1,600 lbs	Success
20 Feb. 1975	Intelsat IV	AC-33	1,600 lbs	Failed–Atlas
22 May 1975	Intelsat IV	AC-35	1,600 lbs	Success
25 Sept. 1975	Intelsat IVA	AC-36	1,820 lbs	Success
29 Jan. 1976	Intelsat IVA	AC-37	1,820 lbs	Success
13 May 1976	Comstar I	AC-38	1,787 lbs	Success
22 July 1976	Comstar I	AC-40	1,787 lbs	Success
26 May 1977	Intelsat IVA	AC-39	1,820 lbs	Success
29 Sept. 1977	Intelsat IVA	AC-43	1,820 lbs	Failed–Atlas
6 Jan. 1978	Intelsat IVA	AC-46	1,820 lbs	Success
9 Feb. 1978	FLTSATCOM F1	AC-44	2,250 lbs	Success
31 Mar. 1978	Intelsat IVA	AC-48	1,820 lbs	Success
29 June 1978	Comstar I	AC-41	1,787 lbs	Success
4 May 1979	FLTSATCOM F2	AC-47	2,250 lbs	Success
17 Jan. 1980	FLTSATCOM F3	AC-49	2,250 lbs	Success
30 Oct. 1980	FLTSATCOM F4	AC-57	2,250 lbs	Success
6 Dec. 1980	Intelsat V	AC-54	2,280 lbs	Success
21 Feb. 1981	Comstar I	AC-42	1,787 lbs	Success
15 Dec. 1981	Intelsat V	AC-55	2,280 lbs	Success
23 May 1981	Intelsat V	AC-56	2,280 lbs	Success
6 Aug. 1981	FLTSATCOM F5	AC-59	2,250 lbs	Failed
4 Mar. 1982	Intelsat V	AC-58	2,280 lbs	Success
28 Sept. 1982	Intelsat V	AC-60	2,280 lbs	Success
19 May 1983	Intelsat V	AC-61	2,280 lbs	Success
9 June 1984	Intelsat V	AC-62	2,280 lbs	Failed–Centaur

[5] Public Law 87-624, 31 August 1962, as found in *Exploring the Unknown, Volume III*, ed. John Logsdon (NASA SP-4407, 1998), 77. See also Hugh R. Slotten, "Satellite Communications, Globalization, and the Cold War," *Technology and Culture* 43 (2002): 315–350.

Date	Mission	Vehicle	Payload Weight to Synchronous Orbit*	Result
22 Mar. 1985	Intelsat VA	AC-63	2,420 lbs	Success
29 June 1985	Intelsat VA	AC-64	2,420 lbs	Success
28 Sept. 1985	Intelsat VA	AC-65	2,420 lbs	Success
4 Dec. 1986	FLTSATCOM F7	AC-66	2,700 lbs	Success
26 Mar. 1987	FLTSATCOM F6	AC-67	2,250 lbs	Failed—lightning
25 Sept. 1989	FLTSATCOM F8	AC-68	2,700 lbs	Success

*The actual weight carried by Centaur to transfer orbit was approximately double these numbers.

Because of the rapid development of satellite technology, communications satellites needed to be redesigned every four to five years.[6] The Delta rocket, manufactured by the Douglas Aircraft Corporation, launched the Intelsat satellites of the 1960s. However, as satellites increased in weight from 85 pounds to 330 pounds, they reached the limits of Delta's ability to lift them. The jump from Intelsat III to IV was a dramatic fivefold increase in weight from 330 pounds to 1,600 pounds sent to geosynchronous orbit. Centaur was the only upper stage available to lift this fourth generation of Intelsat satellites in 1971. By 1985, Atlas-Centaur had launched twenty-four Intelsat satellites with only three failures.

Driven by the popularity of telecommunications, the demand for satellites continued to grow. The weight of the next generation of satellites, Intelsat IVA, increased to 1,820 pounds, with a communications capability twice that of its predecessor. This new Intelsat was launched by an upgraded Atlas-Centaur D-1AR. The new features of Centaur, the result of knowledge gained from the Titan-Centaur missions, included a computer-controlled vent and pressurization system to control tank pressures, new six-pound thrust attitude control engines, and a redundant hydrogen peroxide supply. The payload fairing was redesigned to accommodate the lengthened Intelsat IVA.[7] Both the upgraded Centaur and the new Intelsat worked flawlessly. The next series, Intelsat V, developed by the Ford Aerospace and Communications Corporation, promised an even greater capacity for global communications because of its improved relay capacity.[8]

The commercial success of Intelsat spurred the development of a series of four commercial satellites called Comstar, also launched between 1976 and 1981 on Atlas-Centaur. Roughly the same size and design as the Intelsat IV, Comstar satellites had a longer life because of more efficient solar cells.[9]

---

[6] Martin, "Intelsat V," *Communication Satellites*, 62.

[7] Launch Vehicles Directorate, Lewis Research Center, "AC-36 Flight Data Report," March 1976, Glenn Research Center, Box Flight Reports and Historical Data AC-33 to AC-41, Division Atlas/Centaur Project Office, NASA GRC Records.

[8] For more information on the Intelsat V launches, see "Flight Reports" from Launch Vehicles Division, NASA GRC Records.

[9] See "Flight Reports and Historical Data AC-33 to AC-41" and "Flight Reports AC-42 to AC-49," Division Atlas/Centaur Project Office, NASA GRC Records.

Comsat General, a subsidiary of the Comsat Corporation, owned the Comstar satellites. Comsat leased the satellites to AT&T; this meant that GTE Satellite Corporation (GSAT), a subsidiary of General Telephone and Electronics, could also use them as a relay between their networks of ground stations. AT&T had ground stations in New York, Chicago, San Francisco, and Atlanta, while GSAT served Tampa, Los Angeles, and Hawaii. Comstar provided increased long-distance service in the United States without increasing the number of telephone poles, overhead wires, or microwave towers.[10]

The military also depended on Atlas-Centaur to deliver heavy satellites to orbit. A series of eight satellites, the Fleet Satellite Communications (FLTSATCOM) system, designed and built by TRW Systems, served the space communications network of the Navy, the Air Force, and the Department of Defense.[11] It enabled ships, aircraft, submarines, and ground stations to communicate effectively with each other.

Cost benefits to NASA were significant. Because national policy precluded satellite companies from contracting directly with the launch vehicle manufacturers for launch services, NASA, which owned the vehicles, could charge the makers of commercial satellites for the use of its vehicles, launch facilities, and services. In 1970, the cost of an Atlas-Centaur launch was fourteen million dollars.[12] By 1980, the cost of launching one Comstar satellite had climbed to over twenty-three million dollars. Only about half of this amount came from the launch vehicle hardware. These hardware costs included $4,368,502 for the Centaur stage, $777,058 for the Centaur engines, $2,114,065 for Atlas-Centaur hardware support, $2,563,133 for the Atlas airframe, $1,156,481 for the Atlas engines, $460,701 for guidance hardware, $675,172 for computer hardware, and $305,597 for government-furnished property and services. The other half came from various government support services, including $2,802,724 for contract launch services, $3,852,491 for contract support services, $556,816 for miscellaneous program support, $151,671 for DOD contract administration, $1,519,000 for Air Force range support, $785,322 for NASA project management and engineering, $973,028 for Agency overhead, and $615,000 for use charges.[13] Often it would take NASA up to three years to bill companies for these additional costs. NASA anticipated that it could use revenues from satellite launches to help offset the enormous costs of a Shuttle launch, once all satellite launches were transferred from expendable launch vehicles to the Shuttle.

[10] "Launch Mission Summary Comstar D-3 AC-41," 29 June 1978, Box FLTSATCOM 1978–1983, Division STED, NASA GRC Records.

[11] For more information on the FLTSATCOM launches, see "Flight Reports and Historical Data," NASA GRC Records.

[12] Joan Lisa Bromberg, *NASA and the Space Industry* (Baltimore: The Johns Hopkins University Press, 1999), 111.

[13] Charles J. Tiede to Chief, Launch Vehicles Division, "Final Government Costs for the Third Comstar I Launch," 9 April 1980, Box FLTSATCOM 1978–1983, Division STED, NASA GRC Records.

Old Centaur, New Ariane

The early 1980s proved to be an extraordinarily demanding time for Atlas-Centaur managers as they juggled the competing demands of industry, military, and commercial launches. For example, Lewis engineer James Patterson wrote to NASA Headquarters, "You will note that the Comstar vehicle could be launched 5 February 1981 from Pad A only if Intelsat would be willing to slip their AC-55 launch from 26 February to 5 March 1980, and their AC-58 launch from 8 May to 4 June 1981. The new proposed schedule also takes into account the Navy's request to move up the AC-57 FLTSATCOM launch from 13 November to 28 October 1980."[14]

Despite this heavy use of Atlas-Centaur, Lewis Research Center began to prepare for the termination of all engineering activities associated with its management in anticipation of the Shuttle's taking over its payloads. Termination of the program was forestalled when Intelsat, concerned over delays in Shuttle development, insisted on being able to contract with General Dynamics for four additional Centaurs to be launched prior to 1984. At the same time, NASA, recognizing the limitations of the Shuttle, began to consider using Centaur as an upper stage for the Shuttle (as discussed previously).

This small glimmer of hope in the otherwise gloomy outlook for the Launch Vehicles Division at Lewis permitted Centaur supporters to mount a campaign to keep Atlas-Centaur alive. NASA permitted an Atlas-Centaur Review Board to be convened at Lewis Research Center in October 1980. The blue ribbon committee was composed of representatives from Goddard, Langley, JPL, and the Air Force Space Division. Two significant individuals in the history of Centaur, retired Lewis Center Director Bruce Lundin and former Deputy Associate Administrator for Space Science Vincent Johnson, were called out of retirement to serve as co-chairs. The final report of the committee, submitted February 1981, strongly recommended that Atlas-Centaur be continued as a program into the foreseeable future.[15]

The board emphasized the reliability of Centaur and the soundness of Lewis's management of the program over a period of nearly two decades. The "technical penetration" of Lewis staff included supervision of both General Dynamics, the vehicle's integrating contractor, and associate contractors Honeywell, Teledyne, and Pratt & Whitney. The report recommended against allowing the associate contractors to report to General Dynamics directly because involvement of the Lewis project staff was invaluable: "This supplies both a 'corporate memory' and person-to-person familiarity with their counterparts among contractors."[16] Despite the decade-long threat that the program would be phased out, morale

---

[14] James E. Patterson to NASA Headquarters, "AC-42 Launch Schedule," 5 March 1980, Box FLTSATCOM 1978–1983, Division STED, NASA GRC Records.

[15] *Report of the Atlas/Centaur Review Board*, February 1981, 1, Box 10, NASA GRC Records.

[16] Ibid., 5, Box 10, NASA GRC Records.

at Lewis and among the industry contractors remained high. They were "a true government-industry team" that "supports, stimulates, and challenges the other to produce a whole that [was] greater than the sum of the parts."[17] However, the board warned that Centaur reliability could be affected by a decrease in personnel assigned to Centaur and other cost-cutting measures. It was especially important to keep Lewis Research Center in the picture because of its expertise in software design and verification—an area that General Dynamics had been unable to staff adequately because of imminent phase-out. Another reason was the significant Lewis Research Center role in postflight analysis. Because of their long-term involvement, Lewis engineers had become particularly skilled at uncovering subtle anomalies in test data.

However, there were also significant drawbacks to keeping Atlas-Centaur flying. The board criticized Lewis management for casualness born of familiarity that was evident in preflight checkout. Another shortcoming of the Launch Vehicles Division was its adversarial relationship with Headquarters, a result of the protracted struggle to keep the program from being terminated. The report noted the loss of skilled people both at Lewis and among the contractors. At General Dynamics, the welding of pressure-stabilized tanks for both Atlas and Centaur was a highly specialized skill that few technicians possessed. The same was true of the special expertise needed to manufacture nose fairings and insulation panels. Investigation revealed that an increasing number of "workmanship discrepancies" had occurred since 1978, possibly due to the perception of the pending discontinuation of the program. At Honeywell, technical know-how had declined, along with the reliability of worn-out test equipment. In response to this criticism, the head of the Lewis Launch Vehicles Division wrote, "This situation demonstrates the classical dilemma of a program limping along on small size 'buys.' At this rate our 1960 vintage test equipment will be seeing action cum bailing wire until it dies of exhaustion!"[18]

Even more worrisome was the loss of Pratt & Whitney's expertise in hydrogen/oxygen technology. Staffing at the company had dropped from a level of about 150 engineers in 1967 to just 9. By 1982, the company expected to have only three engineers assigned to the RL10 engine. The technical know-how involved in fabricating thrust chambers, understanding the design and performance margins of the engine, predicting the behavior of cryogenic propellants for cooling thrust chambers, and manufacturing turbopump bearings and injectors could never be replaced.[19] Another area of concern was the anticipated retirement of long-time supervisors, technicians, and inspectors at the Eastern Space and Missiles

---

[17] Ibid., 1, Box 10, NASA GRC Records.

[18] Memo from Larry Ross, Director of Space, to Center Director John F. McCarthy, Jr., 4 June 1981, Box 10, NASA GRC Records.

[19] *Report of the Atlas/Centaur Review Board*, February 1981, 14, NASA GRC Records.

Center at Cape Canaveral. The ground-support infrastructure had been allowed to deteriorate, with worn-out equipment being repaired instead of replaced. The board recommended
that NASA adopt a more flexible system of personnel management that would allow staff
to be shifted between the Shuttle and expendable launch vehicle operations. NASA was
urged to articulate a formal policy stating that it was in the national interest to allow
expendable rockets to coexist with the Space Shuttle, with specific reference to the Atlas-
Centaur launch vehicle as a "necessary element of the U.S. Space Transportation System
into the indefinite future."[20]

It must have come as a relief to the Launch Vehicles Division that a proposed reduction
of the number of engineers from 123 to 65 might be averted. In his formal response, Larry
Ross, Director of the Space Division, outlined steps for implementing the report recommendations. He believed that most of the division's shortcomings noted in the report reflected
a "going-out-of-business" mindset.[21] Predictions of the "last launch" throughout the previous
decade had always proved to be mistaken. Although the future of Centaur remained in
limbo, he recommended that the Launch Vehicles Division go resolutely forward with planning for future launches.

During the same decade that Atlas-Centaur was "limping along" on outdated rocket
technology, a brand-new rocket was being developed by a consortium of European countries. Motivation for financing the building of Ariane differed from country to country; for
the French, it was a political decision driven by French nationalism. France contributed
about 60 percent of the development costs and wanted to be able to launch its own military satellites independently of the United States. The fact that Ariane proved to be an
outstanding commercial success came unexpectedly as a result of NASA's policy blunders.[22]

Ariane, comparable to Atlas-Centaur in power and cost per launch, also had a cryogenic
upper stage. The development of this stage appears to have proved both difficult and
controversial, since the United States regarded cryogenic know-how as an issue of national
security.[23] Despite NASA's skepticism that the Europeans were capable of developing a
launch technology, the European Space Agency pressed forward with development of the
new launch vehicle. When Intelsat purchased a ride for an Intelsat V satellite on Ariane in
1978, it was hailed by the Europeans as a signal "that Europe, notwithstanding the threat of

---

[20] Ibid., 2.

[21] Memo from Larry Ross, Director of Space, to Center Director John F. McCarthy, Jr., 4 June 1981, Box 10, NASA GRC
Records.

[22] John Krige, "The Decision Taken in the Early 1970s To Develop an Expendable European Heavy Satellite Launcher,"
in *A History of the European Space Agency, 1958–1987*, vol. 2 (ESA SP-1235, April 2000), 389–421.

[23] See references in Krige, 413, 421, and L. Sebesta, "The Availability of American Launchers and Europe's Decision 'To
Go It Alone,'" *History of ESA*, vol. 2, 437.

the Shuttle, was now perceived as having a viable, alternative launch service which was going to be competitive on the international launcher market."[24]

The Contradictions of Shuttle Policy

Although President Reagan announced in 1982 that the government would no longer order Titan, Atlas, or Delta launch vehicles, his administration began to work behind the scenes to prepare legislation to commercialize the American expendable launch vehicle industry. An advocate of laissez-faire economics, Reagan championed competition and may have believed that, once liberated from its dependence on government contracting, the expendable launch vehicle industry would be able to win commercial satellite business. The President's Commercial Launch Directive (NSDD 94) set up a new Office of Commercial Space Transportation within the Department of Transportation. Secretary of Transportation Elizabeth Dole organized the Commercial Space Transportation Advisory Committee (COMSTAC) to facilitate communication between the industry and her new office.[25] COMSTAC included representatives of the big three launch vehicle manufacturers—Martin Marietta, General Dynamics, and McDonnell Douglas—and about a dozen startup companies seeking to get into the business of small expendable launch vehicles, the one market not served by the Shuttle. The charge to COMSTAC was to assist the government in drawing up the Commercial Space Launch Act of 1984.[26]

The Space Systems Division at General Dynamics now began to take the first tentative steps toward commercialization. In July 1981, Alan Lovelace, who had served as acting Administrator for NASA during the Carter administration, had become Corporate Vice-president and General Manager of the Space Systems Division. He had great credibility within the aerospace community because of the many years he had spent as head of the Materials Laboratory at Wright-Patterson Air Force Base, Deputy Chief of Staff for R&D for the Air Force Systems Command at Andrews Air Force Base, and then Deputy Assistant Secretary for R&D at the Pentagon. Lovelace recalled that as acting Administrator of NASA, he had strongly believed the Shuttle capable of handling all of NASA's missions, and in this role, he was responsible for informing Lewis Research Center of the imminent cancellation of NASA's Expendable Launch Vehicle Programs.

After he joined General Dynamics, however, Lovelace began to see new opportunities for commercial launch vehicles. In his view, NASA's monopoly of launch services had created a "business of arrogance." A commercial launch service offered customers more control over

---

[24] J. Krige, "The Move from Ariane Development of Production and the Establishment of Arianespace," *History of ESA*, vol. 2, 472.

[25] John M. Logsdon and Craig Reed, "Commercializing Space Transportation," *Exploring the Unknown, Volume IV, Accessing Space* (NASA SP-4407, 1999), 411.

[26] For an insightful account of the entrepreneurial activity stimulated by the new legislation, see Bromberg, 114–131.

their payloads. He strongly believed that if General Dynamics were allowed to sell a launch service directly to the customer, it could manage indirect overhead and become competitive with Ariane.[27] Standing in the way of commercialization was a government contracting environment that stifled initiative. In his view, "the cost-plus mentality had to go."[28]

A 1982 General Dynamics study determined that the company could anticipate an expanding market for commercial rockets. Liberated from dependency on government contracts, the company looked forward to exerting greater control over marketing, design, and production. The study speculated that Atlas-Centaur had the potential to tap 90 percent of this market by upgrading Centaur to reduce costs and increase performance. Systems proposed for modernization included the Centaur avionics, nose fairing, and hydraulics system. The Atlas vernier propulsion system was to be replaced with a simple hydrazine roll control module mounted in the Interstage Adapter between Atlas and Centaur. This change would reduce cost and improve reliability.[29]

In anticipation of commercialization, General Dynamics formed a new commercial entity called CommSpace in March 1983 to market the projected $10-billion international commercial satellite industry. The company promoted the new commercial launch service as a backup launch capability for the Space Shuttle—a capability for which the government would incur no additional costs. General Dynamics offered to assume the risk of a failed launch. It would maintain existing launch facilities in danger of becoming obsolete. The company anticipated that the new commercial initiative would contribute significantly to the U.S. economy.

One of the prime considerations in the move toward commercialization was the possibility of realizing the economies of scale inherent in continuous production. NASA's procurement of Centaur rockets one or two at a time since the end of the Surveyor program had made it difficult to maintain manufacturing expertise.[30] For example, because welding the Atlas and Centaur tanks was highly skilled labor, it took at least three months to get welders retrained after a layoff. Training new workers in the field of electronics was an even greater problem.[31] Quality suffered when well-trained engineers and production workers were spirited away by companies with better long-term prospects.

---

[27] Interview with Alan Lovelace by Virginia Dawson, 19 July 2002.

[28] Ibid.

[29] D. E. Charhut and J. E. Niesley, "The Commercial Centaur Family," paper to 34th Congress of the International Astronautical Federation, IAA-83-233, 10–15 October 1983. See also W. F. Rector III and D. E. Charhut, "The Commercial Centaur Family," Convair Division, General Dynamics Corp., 1983; M. C. Simon and O. Steinbronn, "The Economics of Space Development," General Dynamics, 1984, Box 57, NASA GRC Records.

[30] In the decade of the 1970s, there were only two large contracts: one for nine Centaur stages, dated 24 September 1973, and one for eight stages, dated 8 September 1976. See NASA Historical Data Book, vol. III (NASA SP-4012, 1988), 25–26.

[31] Interview with Stanley Chamberlain by Virginia Dawson, 7 June 2000.

Although the Reagan administration encouraged commercialization in principle, manufacturers of expendable launch vehicles were effectively shut out of the market by the heavily subsidized price of a Shuttle launch once the Shuttle became operational in 1982. As historians John M. Logsdon and Craig Reed have noted, "The reality was that the United States during the 1983–85 period was pursuing two policy goals that were clearly inconsistent."[32] Commercialization was not possible as long as artificially low Shuttle pricing made it impossible for the expendable launch vehicle manufacturers to compete for satellite business.

NASA demonstrated formal compliance with the 1984 Commercial Launch Directive by issuing a request for proposals for commercializing Delta and Atlas-Centaur. Initial agreements with Transpace Carriers (which marketed Thor-Delta for McDonnell Douglas) and General Dynamics were reached in December 1983. They were to go into effect for Delta in 1986 and for Atlas-Centaur in 1987. Also, in September 1984, NASA set up an Office of Commercial Programs focused largely on selling Delta launch services. The budget of this office for its first year was a mere $17.1 million, about 0.2 percent of NASA's total budget.[33]

In the policy debates over the implementation of the 1984 Space Commercialization Act, members of Congress and representatives of the expendable launch vehicle industry argued that Shuttle pricing should at least approximate actual costs. NASA resisted with the rationale that a low Shuttle price was needed to remain competitive with Ariane, which was subsidized by the European governments. The price question pitted the Department of Transportation, which represented the manufacturers of expendable launch vehicles, against both NASA and the satellite makers, who wanted space transportation offered at the lowest possible cost. Cleveland's local paper, the *Plain Dealer*, chided NASA for paying lip-service to space commercialization while working behind the scenes to maintain its monopoly. Since the government contributed between half and three-quarters of the actual cost of a commercial Shuttle launch, expendable launch vehicles could not compete with the Shuttle.[34] As the *Wall Street Journal* reported in March 1985, "The shuttle price question has turned into a noisy bureaucratic trial of strength between NASA and the Transportation Department, and will, in the end, be decided by President Reagan himself."[35] Rudolph Penner, Director of the Congressional Budget Office, predicted that pricing would determine "what gets done in space, how it is done and the fate of competing means of space

---

[32] Logsdon and Reed, "Commercializing Space Transportation," *Exploring the Unknown, Volume IV*, 412.

[33] Joan Lisa Bromberg, *NASA and the Space Industry*, 122.

[34] Patrick Cox, "What Price the Shuttle?" *Plain Dealer* 3 (February 1985): Section B, 1.

[35] Arlen J. Large, "Price Policy on Space Shuttle's Commercial Use Could Launch–or Ground–NASA's Rockets," *Wall Street Journal* (21 March 1985): 54.

transportation."[36] Ariane had a growing market share, and Roland Deschamps, Secretary General of Arianespace, could gloat: "The big error of NASA was to abandon the expendable launch vehicles to private companies and to compete with them through Shuttle."[37]

At the same time, the Department of Defense questioned NASA's insistence that all payloads, both civil and military, be sent aloft on the Shuttle. Concerned that a Shuttle failure might delay sensitive military payloads, the Air Force insisted on keeping at least one expendable launch vehicle manufacturer in business.[38] During congressional hearings, Administrator James Beggs resisted the Air Force initiative, arguing that expendable launch vehicles were obsolete and that support of the expendable launch vehicle industry would undermine the cost-effectiveness of the Shuttle. In the spring of 1984, an Air Force contract for ten upgraded Titan IV rockets was the first indication that at least the Martin Marietta rocket business might survive.[39]

For General Dynamics, the future was more problematic. Prior to 1984, Centaur had enjoyed its most successful record of launches—an impressive string of twenty-eight in a row without a failure. The magic was broken when the first flight of an improved Atlas G (AC-62, 81 inches longer than the previous Atlas), mated to an upgraded Centaur vehicle, failed.[40] The failure investigation kept the next three Atlas-Centaur launches on hold. Investigations completed in August 1985 concluded that the failure was caused by a liquid-oxygen leak through a 4-inch crack in the intermediate bulkhead, which occurred during Centaur's separation from the booster.[41] The company could ill afford this interruption in service because, by this time, Ariane had gained an impressive 30 percent of the world market for satellite launchers.[42]

---

[36] Ibid.

[37] "NASA Does Not Admit Its Error in Putting All Eggs on Shuttle," *Interspace: The European Satellite & Space News* (11 January 1985).

[38] The role of the Air Force is discussed by Craig Reed in "Factors Affecting U.S. Commercial Space Launch Industry Competitiveness," *Business and Economic History* 27 (1998): 229; Ragsdale, 150.

[39] See discussion by Joan Lisa Bromberg, 131–132. See also previous note. (The Air Force ordered another twenty-five Titan IV vehicles at a total cost of more than $4 billion after the *Challenger* accident.)

[40] Mary Ann Peto and Roland Raab, "Stretched Atlas-Centaur Will Boost Payload Capacity," 3 May 1984, NASA Historical Reference Collection, Washington, DC. Improvements to the new "silver-throated Centaur" included a pressure-fed system to replace the boost pumps, a new hydrazine attitude-control rocket system, and silver plating on the interior of the rocket nozzle throat to increase thrust. The result was an extremely powerful unpiloted vehicle with the capability of placing an extra 500 pounds into geosynchronous orbit.

[41] General Dynamics, "Atlas-Centaur AC-62 Failure Investigation Interim Report," Report No. GDC-SP-84-045, November 1984, Box AC-62 Failure Investigation, Division Atlas-Centaur Project Office, NASA GRC Records; General Dynamics, "Atlas-Centaur AC-62 Failure Investigation Final Report," Report No. GDC-SP-85-018, August 1985, Box AC-62 Failure Investigation, Division Atlas-Centaur Project Office, NASA GRC Records.

[42] J. Krige, "The Move from Ariane Development of Production and the Establishment of Arianespace," *History of ESA*, vol. 2, 481.

A suit by Transpace Carriers, Inc. (which marketed Delta for McDonnell Douglas), claimed that Arianespace was unfairly and illegally engaging in "predatory pricing" of launch services. This attack on Ariane pricing was indirectly an attack on Shuttle pricing, since both were heavily government-subsidized. Even after raising the price of a Shuttle launch from $38 million to $71 million, this price was still far below the actual cost of a launch. "U.S. pricing policy on launchers in 1984/5," historian John Krige wrote, "was thus torn between the fundamental contradiction of trying to protect the Shuttle from Ariane while encouraging private industry to develop a viable commercial ELV service in parallel."[43]

After *Challenger*

The *Challenger* disaster in January 1986 precipitated a complete reversal of Shuttle policy and opened the way for the development of a viable expendable launch vehicle industry in the United States. In December 1986, President Reagan issued National Security Decision Directive 254, stipulating that NASA would no longer launch satellites for private companies or foreign governments, except those that required a human presence in space or involved national security. This decision took the Shuttle out of competition with expendable launch vehicles for revenue-generating business.

Analysis of the Shuttle's manifest after the loss of *Challenger* showed that of a total of 186 payloads scheduled for launch between 1986 and 1989, 103 were appropriate to be flown on expendable launch vehicles. Lee R. Scherer, director of the commercial space systems division for General Dynamics, pointed out that the Shuttle pricing policy had precipitated the closing down of the production lines for Delta, Atlas-Centaur, and Titan 34D rockets. Now the hiatus in Shuttle flights left commercial satellite customers with only one option—the Ariane rocket. He concluded that the *Challenger* tragedy had created many uncertainties, but one thing was clear: "Our country must never again allow itself to get into the current posture where we have been completely dependent on a single form of space transportation."[44]

NASA's Mixed Fleet Study, completed in May 1987, complemented the new policy. It emphasized that in addition to the Shuttle, the nation needed expendable launch vehicles of different sizes and capabilities. Richard Truly, Associate Administrator for Space Flight, parceled out the management of NASA's expendable launch vehicles among NASA Centers. Lewis Research Center received the management of intermediate- and large-class expendable launch vehicles, namely Atlas-Centaur and Titan-Centaur. Goddard took over the management of small- and medium-class expendables, such as the Delta vehicle, as well as payload

---

[43] J. Krige, 483.

[44] L. R. Scherer, "Expendable Launch Vehicles as a Complement to the Space Shuttle," paper presented to the AIAA Communications Satellite Systems Conference, March 1986. See also Keith Mordoff, "ELV Makers Gear for Production Restart," *Commercial Space* (Spring 1986): 44–47

processing at Wallops Flight Facility. Kennedy Space Center was assigned the assessment of expendable launch vehicle processing and launch countdown for launches involving NASA.[45]

The *Challenger* disaster not only changed NASA policy, but also raised questions over NASA's managerial competence. While the problems related to the *Challenger* tragedy received wide media coverage, there were Agency concerns for NASA's management of expendable launch vehicles. An Atlas-Centaur carrying a FLTSATCOM was struck by lightning during a driving rainstorm on 26 March 1987. Forty-eight seconds into launch, the Centaur computer ordered an incorrect yaw steering command that caused the rocket and its payload to explode 50 seconds later. This failure in the wake of *Challenger* precipitated a congressional investigation. NASA revealed that the lightning strike had caused the computer to issue an erroneous command.[46] But the failure involved more than just the weather. The Centaur team at Lewis Research Center was reprimanded for work practices that exhibited "inadequate emphasis on procedural rigor and discipline."[47] The Centaur team admitted that the obvious indication of a lightning hazard "escaped the launch management team because of imprecise communications, lack of awareness, or both."[48] Just six months before this accident, another NASA review board had questioned procedures for making decisions that related to weather at Cape Canaveral.[49] Had the recommendations from this report been followed, NASA would have waited for better weather to launch AC-67. Political cartoons satirized NASA weather forecasting, and one caption stated that even Ben Franklin would not have flown his kite during such lightning conditions. An editorial in *Aviation Week & Space Technology* described the AC-67 accident in the context of the previous year's *Challenger* tragedy. The editor wondered, "What other time bombs are ticking away in the agency's hidebound management structure that need to be rooted out to prevent further loss of life, limb, and property?"[50]

The answer came four months later. During preparation for the last scheduled NASA launch of an Atlas-Centaur, a workstand bounced against the empty but pressurized hydrogen tank. It sliced a gash in the side of the tank, causing the vessel to rip violently

---

[45] "Field Center Expendable Launch Vehicle (ELV) Roles, Responsibilities, and Organizational Structure," 17 December 1987, Box 57, NASA GRC Records.

[46] General Dynamics, "Atlas-Centaur Postflight Analysis AC-67," Report No. GDSS-SP-87-011, July 1987, and "Determination, Findings, Observations, and Recommendations," report given before Congress, 30 June 1987, Box AC-67 Failure Investigation, Division 6500 John Gibb, NASA GRC Records.

[47] "Atlas-Centaur 68 Investigation Board Report on the 13 July 1987 Centaur Liquid Hydrogen Tank Mishap," Box AC-68, Division Atlas-Centaur Project Office, NASA GRC Records.

[48] Ibid.

[49] Craig Covault, "Atlas Accident Inquiry Board Finds Violation of Launch Commit Criteria," *Aviation Week & Space Technology* (16 May 1987): 25.

[50] "Atlas Launch Follies," *Aviation Week & Space Technology* (18 May 1987): 11.

apart. Several workers sustained minor injuries. Investigation revealed a lack of discipline on the part of the workers at Cape Canaveral, again attributed to poor NASA management.[51]

NASA was also criticized for the inefficiency of payload processing at the Cape and for allowing the launch infrastructure to deteriorate. Investigation revealed that about half of the seventy-seven expendable launch vehicle payloads launched between January 1977 and February 1988 had experienced delays of an average of two weeks. Sixty-five percent of the delays were caused by payload or launch vehicle problems, with nearly 30 percent of the payloads requiring a storage period.[52] It took longer to process vehicles for launch because the Air Force controlled the launch pads, while the ground-support equipment, including mission control and the tracking stations, were NASA's responsibility. The expendable launch vehicle manufacturers and their satellite customers had to contend with NASA rules, practices, and priorities, as well as those of the Air Force. Arianespace had the advantage of a streamlined launch facility without any kind of government oversight.

To address these and other problems, the Air Force drew up a Model Range Use Agreement and sought ways to keep the expendable launch vehicle manufacturers in business. In 1984, to ensure that the military had access to space, the Air Force had contracted with Martin Marietta for the ten Titan IV boosters. After *Challenger*, the Air Force increased this order to twenty-three vehicles at a cost of $5.1 billion. Secretary of the Air Force Edward (Pete) Aldridge wanted to end Air Force dependency on NASA by creating a robust commercial industry. The new contract permitted Martin Marietta to market a commercial version of Titan III to ten satellite customers. In the spring of 1987, the Air Force contracted with McDonnell Douglas for eleven Delta II medium launch vehicles with an option for twenty additional vehicles. At this point, of the three established manufacturers in the expendable launch vehicle business, only General Dynamics lacked a significant new contract with the Air Force.[53]

Despite the loss of the medium launch vehicle competition to McDonnell Douglas, Alan Lovelace forged ahead with a major financial commitment to commercialization. The cancellation of Shuttle/Centaur after *Challenger* left General Dynamics with a cadre of experienced Centaur people to initiate the commercialization effort.[54] Lovelace convinced the company's board of directors to finance the manufacture of eighteen Atlas-Centaur vehicles before they actually had orders for them. The company granted the Space Systems Division $125 million to start up the assembly line.

---

[51] "Atlas-Centaur 68 Investigation Board Report on the 13 July 1987 Centaur Liquid Hydrogen Tank Mishap," Box AC-68, Division Atlas-Centaur Project Office, NASA GRC Records.

[52] Creighton A. Terhune and Shirley P. Green, "Payload Processing Study for ELV Launches," paper presented at the 26th Space Congress, 1989, AEDC Tech Library, Arnold AFB, TN.

[53] Ramon L. Lopez and Greg Waskul, "New Life for Expendable Launchers," *Space Markets* (Spring 1987): 4.

[54] Interview with Marty Winkler by Virginia Dawson, 21 March 2001.

In the fall of 1986, an Air Force Request for Proposal for twelve medium launch vehi-cles once more pitted the three major expendable launch vehicle manufacturers, McDonnell Douglas, Martin Marietta, and General Dynamics, against each other. The competition for this Air Force contract created a make-or-break situation for General Dynamics. The second Medium Launch Vehicle (MLV II) contract would give an edge to the winner over its competitors because the military order would allow the company to leverage the Air Force contract against the startup costs of commercial production.[55] At this time, at the urging of the Air Force, General Dynamics dropped the name Centaur when referring to its commer-cial vehicle.

In May 1988, Lovelace, who presided over the American Institute of Aeronautics and Astronautics (AIAA) that year, found a note from Pete Aldridge on his plate when he returned to his seat on the dais during the annual banquet. It simply said, "Congratulations, you have won the MLV II competition." This welcome news vindicated Lovelace's auda-cious decision to start up production at company expense. The $340-million Air Force contract for ten launch vehicles represented an important step toward underwriting commercialization of the entire American expendable launch vehicle fleet. With commer-cialization, the contractor, not the government, was responsible for the integration of the launch process. To make sure everyone in the company focused on manufacturing a quality product, General Dynamics planned to self-insure the vehicles, offering customers a free ride should the booster fail. "This commitment to a program with a value of more than $1 billion will create over 6,000 jobs for our domestic work force," Lovelace stated in congres-sional testimony on 5 October 1987.[56]

Because of his NASA background, Lovelace understood the cultural changes NASA needed to make in order to transition from purchasing hardware to purchasing launch ser-vices, but he was convinced that commercialization would result not only in greater reliability, but also in significant cost savings for NASA. The company planned to offer launch services to customers, including NASA, at a fixed price. Lovelace insisted that it was in the best interest of the government for NASA to become a customer of American compa-nies in the launch vehicle business. He stated to Congress, "Advantages to the United States Government in procuring launch services from a robust and competitive stable of domestic commercial launch vehicles are numerous. Effectively the United States Government can enjoy the advantages of a multiyear procurement and quantity buys with only limited financial

---

[55] See Theresa Foley, "U.S. Manufacturers Begin the Job of Rebuilding the U.S. Space Program: ELVs," *Commercial Space* (Fall 1986): 16–21.

[56] U.S. Senate Committee on Commerce, Science, and Transportation, Subcommittee on Science, Technology, and Space, 5 October 1987, 60.

obligation and risk. Attendant lower cost, flexibility in scheduling, and resource management can all be realized."[57]

NASA, however, resisted the idea of contracting for a launch service. The Commerce Department, which operated the Geosynchronous Operational Environmental Satellite (GOES) weather satellites, forced the issue. It insisted that NASA use a commercial vendor to launch the three GOES weather satellites and, when NASA resisted, simply threatened to find another procurement agent. GOES became NASA's first commercial contract and represented a milestone in the effort to commercialize the launch vehicle industry.[58]

The Barter Agreement

One of the most important elements of the GOES contract was the successful negotiation of a barter agreement to turn over government property and equipment in exchange for launch services.[59] Government property involved in this agreement included all NASA-owned equipment—gantries, storage tanks, bunkers, etc—at Launch Pads 36A and B and test facilities, as well as equipment owned by NASA and located on contractor property. Crafting the agreement took enormous contracting creativity on the part of procurement officers at Lewis because there was no precedent for turning over government property to the private sector at fair market value rather than at cost. The launch pads were not part of the barter agreement because they were still owned by the Air Force, which leased them to General Dynamics. But complicating the determination of value of NASA-owned facilities and equipment at the Cape was the fact that they had been allowed to deteriorate in anticipation of cancellation of the program. They were worth far less than the government's initial investment. Another complicating factor was the unusual method of Lewis contracting. When Silverstein had taken over the Centaur program in 1962, he had insisted that General Dynamics be treated as an associate contractor. Contractors for subsystems like Pratt & Whitney and Honeywell had all contracted directly with Lewis. By keeping Lewis at the center of all aspects of vehicle management, the government was able to control costs and leverage its considerable technical expertise. However, this model for contracting made crafting the barter agreement more complicated. In 1987, instead of managing a single contract with General Dynamics as a prime contractor, Lewis was managing about fifty-nine major contracts, valued in excess of $1.5 billion.

The termination of the Shuttle/Centaur program in 1986 and the commercialization of Atlas-Centaur called for the disposition of about $360 million in government-owned prop-

[57] Ibid., 62.

[58] Joan Lisa Bromberg, *NASA and the Space Industry*, 158–159.

[59] Interview with Ronald Everett by Virginia Dawson, 17 December 2001.

erty that remained in contractor hands.[60] This included vehicle components, special test equipment, tooling, ground-support equipment, raw materials, and partially finished work at the time of cancellation. NASA had to find a way to place a value on the government property involved in these contracts. Once General Dynamics owned this property, the company could legally offer the entire Atlas-Centaur package as a launch service to commercial satellite customers and NASA.

General Dynamics agreed to rehabilitate and provide launch services for AC-68, damaged extensively in July 1987 by the careless workstand accident. The company also agreed to supply the entire vehicle and launch services for the Combined Release and Radiation Effects Satellite (CRRES) mission (AC-69), originally scheduled for flight on the Shuttle. In exchange, General Dynamics received the government-owned property from Honeywell, Pratt & Whitney, Teledyne, and Rocketdyne, as well as the legal rights to NASA-owned property at company headquarters and at the Cape. It also received rights to the vehicle design. Thanks to the barter agreement, General Dynamics could proceed with the upgrade of the launch pads and changes to the vehicle design to lower costs and increase reliability. This was a major financial undertaking, but one that General Dynamics was willing to finance. One of the company's immediate priorities was to reduce the inefficiencies that had crept into launch practices over a period of thirty years.

A significant problem remained. The loss of AC-68 was not only costly, but it also came at a point when many NASA engineers with critical experience had been reassigned to other projects. John Gibb, manager of the Centaur project office at Lewis, was assigned the daunting task of supervising the rebuilding of the last NASA-owned Centaur. For General Dynamics, it was an equally challenging undertaking that involved restarting the Centaur assembly line after it had been shut down for several years. Because of the production hiatus and the loss of key personnel, NASA noted that the company "had difficulty in recapturing the formula" for manufacturing Centaur.[61] After two years, Gibb was able to accept delivery of a new rocket. On 25 September 1989, the FLTSATCOM satellite was placed into orbit by AC-68.[62] James Womack, Director of Expendable Vehicles at the Kennedy Space Center, congratulated Gibb with appreciation mixed with relief when a week "plagued with uncertainties" ended with a flawless launch.[63]

[60] Memo, "Report on Audit of the Disposition of Atlas and Centaur Property," A-LE-88-002, 18 November 1988, part of the Audit Report, Office of Inspector General, files of Ronald Everett, NASA GRC.

[61] Jean E. Klick to John Klineberg, 29 September 1989, Box AC-68 Flight Reports and Historical Data, Division Atlas-Centaur Project Office, NASA GRC Records.

[62] "Atlas-Centaur Postflight Analysis AC-68," Box AC-68 Flight Reports and Historical Data, Division Atlas-Centaur Project Office, NASA GRC Records.

[63] John W. Gibb to James L. Womack, 5 October 1989, Box AC-68 Flight Reports and Historical Data, Division Atlas-Centaur Project Office, NASA GRC Records.

Headlines in the *Washington Times* noted, "Nostalgic NASA launches last unmanned payload."[64] After 448 launches, dating back to the Thor-Able on 11 October 1958, the Agency ended its management of expendable launch vehicle launches. From this point on, NASA would have to purchase launch services from its former contractors. William B. Lenoir, NASA's acting Associate Administrator for spaceflight, officially commemorated the end of the old way of doing business in a letter to Lewis Director John M. Klineberg. Lenoir, a former astronaut, was responsible for developing, operating, and implementing all policy for the Space Shuttle and all U.S. government civil launch activities. Lenoir congratulated John Gibb, Vernon Weyers, and the Lewis team for "exemplary management of the mission."[65] The success of the last Centaur mission, he said, was indicative of the outstanding cooperation between government and industry that had been the hallmark of the Centaur program.

Congress, which saw commercialization as a means to reduce the enormous federal deficit and improve the balance of trade, fully supported the barter agreement. Amendments in 1988 to the 1984 space commercialization act added guidelines for establishing insurance limits for loss or damage to government property, as well as U.S. government indemnification for third-party liability above $500 million, or what was reasonably available. The purpose of these amendments was to provide stability and support for a fledgling commercial space industry. Because one launch cost between $50 and $100 million, Congress focused on the fact that a viable industry had the potential to affect the United States balance of trade significantly. Bill Nelson, Chairman of the Subcommittee on Space Science and Applications, stressed that nurturing this industry was in the national interest. "The sale of one commercial launch by a United States company is equivalent to the import of ten thousand Toyotas,"[66] he said on more than one occasion.

Nelson was not satisfied with NASA's apparent lack of effort to simplify its procurement guidelines. In a letter to Dale D. Myers, acting Administrator, on 3 May 1989, Nelson asked why NASA seemed unwilling to work toward the modification of the Federal Acquisitions Regulations in order to expedite NASA procurement. Most of these guidelines were not compatible with a fixed-price contract. "It is time that NASA recognized that it is not the only participant in our space program," he said acerbically. "This nation has a mature commercial launch industry that is fully capable of meeting the government's launch requirement."[67]

---

[64] "Nostalgic NASA Launches Last Unmanned Payload," *Washington Times* (26 September 1989).

[65] William B. Lenoir to John M. Klineberg, 10 October 1989, Box AC-68 Flight Reports and Historical Data, Division Atlas-Centaur Project Office, NASA GRC Records.

[66] "Summary of November 9, 1989, House Subcommittee Hearing on Commercial Space Launch Activities," 27 November 1989, Box 57, DEB vault. See also discussion by Craig Russell Reed, "U.S. Commercial Space Launch Policy Implementation, 1986–1992" (Ph.D. dissertation, George Washington University, 1998), 21–24.

[67] Bill Nelson to Dale Myers, 3 May 1989, Box 57, NASA GRC Records.

Asked to respond to this critique, Centaur staff at Lewis Research Center took issue with the premise that the nation had a mature commercial launch vehicle industry, noting that although there were at least thirteen firms that wanted to get into the business, there were only three firms (General Dynamics, McDonnell Douglas, and Martin Marietta) with a demonstrated technical capability to provide launch services. Although NASA was attempting to "widen as far as possible the gate of opportunity for all domestic launch service firms,"[68] the Agency needed to balance its different obligations to itself, to industry, and to the taxpayer. The issue was the degree to which NASA should risk expensive payloads on commercial launch vehicles. "Simplistically, the easiest approach from NASA's perspective, and the most purely 'commercial' would be to limit awards to only those pre-qualified firms with a proven track record of successful performance and not risk NASA payloads on unproven designs."[69] The Lewis staff emphasized that NASA payloads were "unique, one-of-a-kind, high cost, virtually irreplaceable spacecraft" that required unique launch services. In contrast, commercial launches were "more repetitive, less unique, and far less expensive to replace." They cited the successful negotiation of the first NASA contract for General Dynamics launch services in 1988. During negotiations with General Dynamics for GOES for the National Oceanic and Atmospheric Administration, they were able to waive the certified cost and pricing data requirement. They believed that extreme caution and insistence on NASA oversight were justified. "While we have dramatically scaled back our oversight requirements wherever possible, prudence dictates that we cannot blindly accept assertions that all is being done properly."[70]

Members of COMSTAC pushed NASA to accept the new commercial mandate. For example, Dennis Dunbar, Vice President of Programs and Technical Operations for General Dynamics Commercial Launch Services, Inc., pointed out in testimony before COMSTAC's Subcommittee on Procurement that commercialization entailed "a complete change of mind set on the part of those doing the procuring for the government."[71] To procure only one vehicle at a time could never be cost-effective. There were distinct advantages to the government in procuring launch vehicles in blocks with common specifications. Over the previous decade, the government had contracted for an average of one and a half vehicles per year. With the approach advocated by General Dynamics of eight vehicles per year, costs were projected to be reduced by 64 percent.

---

[68] "Talking Points on ELV Commercialization," in draft of response to Bill Nelson, 22 May 1989, Box 57, NASA GRC Records.

[69] Ibid.

[70] Ibid.

[71] "Testimony of COMSTAC Subcommittee on Procurement to Subcommittee on Space Science and Applications on Implementation of the Commercial Space Launch Act and its Amendments and Space Transportations Services Purchase Act of 1989," 9 November 1989, ELV Program Planning, 1986–1989, NASA GRC Records.

Dunbar compared the General Dynamics first commercial contract with Eutelsat, a French satellite company, with the GOES contract. The Eutelsat contract was 91 pages, compared to 4,250 pages for the NASA GOES proposal. Nevertheless, Dunbar thought that the NASA contract represented significant progress. "After spirited negotiations, GD was able to convince the government that many of their usual requirements just did not fit in a commercial buy. There were many arguments—we won some and lost some—but there was healthy movement in what we consider the right direction."[72] Subsequent requests for proposals were answered in an average of 500 pages, a great step forward for the industry and NASA. The principal contracting officer on the NASA side of these negotiations, Thomas D. Tokmenko, incorporated numerous unique provisions that became a model not only for buying commercial launch services, but also for the purchase of commercial satellites by the government.

Officers of General Dynamics, however, continued to press the Office of Federal Procurement Policy to simplify the Federal Acquisition Regulations. To facilitate this process, the COMSTAC subcommittee presented the government with a model request for proposals (RFP) and model contract to assist in the procurement process. Some of the recommendations included a fixed payment schedule rather than progress payments. This change was needed because the manufacturer now produced launch vehicles "on speculation, at contractor risk, and not allocated to specific missions until just before completion."[73] Second, since the contract was for launch services, not hardware, the statement of work was limited to performance requirements such as a peculiar orbit requirement, interfaces, and environments. The government no longer had the right to dictate how a contractor achieved that performance. Third, for the first time, the contractor could insist that the government adhere to a strict schedule with penalties for unreasonable delays. The contractor was liable for costs associated with termination. To allow the contractor to make a profit, the government permitted pricing to be considered independently of the cost of a particular launch.

As a condition for bidding the MLV II contract, Lovelace succeeded in persuading the directors of General Dynamics to increase planned production from eighteen to sixty-two vehicles. This increase was again internally financed. In fact, since the restart of the company's factory, Al Lovelace had asked the company's board of directors for funding in increments of $100 million so often that $100 million became known as the "Lovelace unit." Fortunately, this significant investment began to show results. The next year, in addition to contracts for Intelsat VII and Navy ultra-high-frequency communications satellites,

[72] Ibid.

[73] Ibid.

General Dynamics won business from the Italian Space Agency for its Satellite per Astronomia X (SAX) and from NASA to launch the Solar and Heliospheric Observatory (SOHO) mission in August 1989. It was fortuitous for American companies that Arianespace experienced a failure of an Ariane II, carrying an Intelsat V satellite, the same year as the *Challenger* disaster. This worldwide hiatus in commercial launch services allowed General Dynamics and Transpace Carriers, Inc., an entrepreneurial company set up to market the McDonnell Douglas Delta rocket, to break into a market dominated by Ariane.[74]

An Evolving Family of Launch Vehicles

In the days when NASA had procured launch vehicles from General Dynamics one or two at a time, NASA had supplied the engines and the Centaur guidance system and computer as government-furnished equipment. In the new world of commercialization, the company took responsibility for procurement. Roy Roberts commented that for the first time "people really understood now what NASA was bringing to the table."[75] Faced with Ariane's daunting head start in commercialization, General Dynamics engineers dusted off NASA studies for Centaur improvements never implemented because of the "going out of business" mentality. The company could no longer expect the government to finance development costs. However, contrary to the expectation that innovations would cease once NASA no longer financed improvements to Centaur, commercialization actually promoted innovations.

To make Atlas-Centaur more appealing to the satellite market, General Dynamics began to think in terms of a family of rockets that could be swiftly adapted to different customer needs. Within six years of the decision to restart Atlas-Centaur production, the company had developed four new designs: Atlas I, Atlas II, Atlas IIA, and Atlas IIAS. (These were all Atlas-Centaur vehicles, but the company dropped the hyphenated name in favor of Atlas.) Each represented an advance in performance, reliability, and payload envelopes, and reduced the time it took to manufacture and launch the vehicle. All four models were included in the sixty-two vehicles built at company expense. Atlas-Centaur engineers emphasized that each successive model had "evolved" from the previous one.

Building more than one or two vehicles at a time allowed the company to formulate a strategic plan that included continuous improvements to the vehicle. This strategy required an enormous investment on the part of the company. The philosophy of continuous improvement encouraged greater standardization, which in turn reduced costs and increased reliability. The Atlas I, actually an improved Atlas G-Centaur D-1A, became General Dynamics's first commer-

---

[74] Craig Russell Reed, "U.S. Commercial Space Launch Policy Implementation, 1986–1992" (Ph.D. dissertation, George Washington University, 1998), 306.

[75] Interview with Roy Roberts by Virginia Dawson, 21 March 2001.

cial launch vehicle. A pressure-fed system replaced the Centaur boost pumps, an improvement whose feasibility had been demonstrated by NASA years before. This change reduced complexity and cost. The avionics system was upgraded. Because it was less expensive to manufacture than a composite nose fairing, a new 14-foot-diameter metal nose fairing was developed to accommodate larger payloads originally intended for the Shuttle or Ariane. In the Atlas booster, a computer-controlled pressurization system with redundant sensors replaced the old mechanical regulators. This innovation made the system not only more reliable, but also more versatile.

Atlas II represented an important evolutionary step. One key change in Atlas II and all later versions was a fixed foam insulation system glued to the outside of Centaur's hydrogen tank. This eliminated the mechanical complexity and weight of jettisonable insulation panels. Although the company had been working on the concept of foam insulation for almost a decade using NASA contracted research and development funds, there was no incentive to implement these changes under the old system. Up to this point, conventional wisdom in the company dictated that change involved compromise. It was thought that increased reliability would affect cost and performance. Edward Bock recalled that during the years when Atlas-Centaur vehicles were being produced one or two at a time, "the message was 'don't make changes.'"[76] The switch to foam insulation, however, not only increased performance and reliability, but also reduced cost. As a result of successful implementation, the new philosophy of continuous improvement permeated the company.

A completely new solid-state avionics suite replaced the obsolete guidance and navigation system. The new avionics system, also the fruit of NASA contracted research and development funds sponsored by Lewis Research Center in the 1970s, increased computing capability, reliability, and redundancy while it simultaneously reduced cost from approximately $8 million per vehicle to $2 million.[77] At the same time, the company contracted with Rocketdyne to upgrade the Atlas engine system to provide greater thrust. Atlas and Centaur were stretched 9 feet and 3 feet, respectively. The first Atlas II launched Eutelsat II into geosynchronous transfer orbit in December 1991.

Quality control and on-time delivery improved dramatically after General Dynamics established a mentoring program to encourage strong relationships with engineering peers in supplier companies. Marty Winkler said, "The production line wound up being a real production line rather than a haphazard collection of things to do."[78] Suppliers learned in far greater detail how their particular part was going to be used in the vehicle, and they began to take responsibility that went beyond the specified "form, fit, and function."

[76] Interview with Edward Bock by Virginia Dawson, 22 March 2001.

[77] Interview with Martin Winkler by Virginia Dawson, 21 March 2001.

[78] Ibid.

Atlas-Centaur 69, the first commercial launch of a NASA mission. The CRRES satellite, originally scheduled for launch from the Shuttle, was part of the barter agreement that paved the way for commercialization of Atlas-Centaur. (NASA C-91-02792)

Greater standardization meant that the vehicle no longer had to be virtually taken apart, tested, and rebuilt at the Cape. This improvement had an enormous impact on how long it took to prepare for launch. Management of the launch manifest changed significantly with commercialization. In the NASA era, launches were scheduled years in advance and General Dynamics planned no more than three or four launches a year. After commercialization, the number of launches averaged to 6.36 per year, with a peak number of launches occurring in 1995 when 11 Atlas-Centaur vehicles were launched. Even more important than the number of launches was the new interaction between manufacturer and customer that commercialization allowed for the first time. The manifest became dynamic—driven by customers. The company was able to reduce the time between the date of signing a contract and actual launch from thirty months to about six. The need to be responsive to many different customers demanded greater flexibility. Roy Roberts pointed out that because a commercial customer regards an Atlas-Centaur simply as transportation, "they expect you to be ready, to go when *they* want to go and be successful."[79]

For the next iteration, the Atlas IIA, Pratt & Whitney redesigned the RL10 to increase thrust and added an extendable nozzle. The added length allowed the rocket to be more efficient at higher altitudes by holding in the exhaust plume, rather than allowing it to expand rapidly. The Atlas IIAS featured four solid rocket boosters to increase overall booster capability. This was a major design change that the company had resisted for many years because of the belief that a pressure-stabilized tank was too delicate to support solid rocket boosters. AC-69, the first commercial launch of a NASA mission, carried the Combined Release and Radiation Effects Satellite (CRRES) in July 1990. For the first time, no NASA official sat in the blockhouse at the Cape during the countdown. Even more startling, "General Dynamics" was now emblazoned down the side of the rocket where "United States" had always been proudly displayed before. It was indeed the end of an era.

Continuity and Change

Despite an enormous General Dynamics investment in commercialization, it was not clear by the early 1990s whether the company's space division could survive. Most of the startup companies in the launch vehicle business had already failed, but even the future of the three established manufacturers looked doubtful. NASA engineers Nieberding and Spurlock argued that expendable launch vehicles should be considered a national resource. In an article for *Space Commerce* in 1990, they emphasized that although the government now purchased launch services, NASA missions continued to be extremely expensive because NASA payloads were unique and required large-scale government involvement. They pointed out that when Atlas-Centaur was developed in the 1960s, reliability, not low cost, was the criterion. "From our experience with Atlas/Centaur and Titan/Centaur, these systems are unforgiving and require scrupulous, expensive attention to detail to achieve a high probability of success."[80] They asserted the value of continued government oversight of unique and costly NASA missions. They questioned whether the industry could afford the large investment required to develop a reliable, low-cost system, particularly when they were faced with government-subsidized foreign competition. Fierce international competition might drive American companies out of the expendable launch vehicle business, cutting off United States government access to space.

Continued government involvement, however, made it more difficult for U.S. companies to compete with Ariane. Bruce Berkowitz, a policy analyst, pointed out that the expense and rigidity of specification systems used by NASA and the Air Force to ensure quality tended to discourage innovation. Government specifications controlled the design of vehicles built for the commercial market because it was not economical to manufacture different models for different customers. The use of the same launch facilities for both government and commercial payloads also complicated operations and caused delays. NASA and Air Force missions took precedence over commercial payloads, while Ariane's manifest put the commercial customer first. Berkowitz found it ironic that a country that extolled the benefits of a free market system allowed the U.S. government to call the shots. "Unless we adopt policies that unleash the potential of the private sector," he warned, "we are using only one leg to catch up in the race."[81]

Of the three big manufacturers, the future of General Dynamics was the most problematic because of its large investment in commercialization. In 1990, the Atlas program lost $300 million. During the next two years, the company lost more revenues when three Atlas I launches of commercial payloads failed (AC-70, -71, and -74). These failures made it extremely difficult for the company to sell its services to satellite customers. Two of these failures were attributed to the icing of a valve in the Pratt & Whitney RL10 engine. The valve was required when boost pumps were eliminated in 1984 and worked well for several flights. Then a slight reduction in valve clearance caused the valve to stick. The third loss of an Atlas-Centaur was caused by the flawed design of a small screw in the Atlas Rocketdyne engine. The faulty screw had been used without incident for thirty years prior to this failure, whose timing was especially unfortunate.[82]

After the redesign of the screw, the company launched Telstar-401 into geosynchronous transfer orbit in December 1993. The success of this launch did little to dispel the gloom

---

[80] J. J. Nieberding and O. F. Spurlock, "U.S. ELVs: A Perspective on Their Past and Future," *Space Commerce* 1 (1990): 32.

[81] Bruce D. Berkowitz, "Energizing the Space Launch Industry," *Issues in Science and Technology* (Winter 1989–1990): 77–83.

[82] Interview with Marty Winkler by Virginia Dawson, 21 March 2001. See also "Builder Suspects Engine Failure in Failed Liftoff," *New York Times* (20 April 1991).

that had settled over the company. By this time, the company had launched eighty-one Atlas-Centaur vehicles with a success rate of over 93 percent, but three failures in twenty-three months sealed the fate of the company.[83] To the General Dynamics corporate headquarters, it appeared that the enormous investment in upgrading the launch pads, incorporating new technology into the launch vehicle, and starting up the assembly line could never be recovered. Former Shuttle commander turned insurance company executive Frederick Hauck recalled that at this time, Atlas was virtually uninsurable.[84] A three-year hiatus in sales resulted in large losses for the company and prompted the decision to sell the Space Systems Division of the company. A new chief executive officer, Michael Wynne, was brought in to prepare the company for sale.

In March 1994, Martin Marietta, manufacturer of Titan, bought the Space Systems Division of General Dynamics for $208.5 million. This was considered a bargain price for a major company division with a backlog of twenty-nine Atlas-Centaur contracts and fifteen upper-stage Titan-Centaur contracts through the end of the decade. The chairman and chief executive of Martin Marietta, Norman Augustine, commented, "There just isn't enough work to be done to justify all the companies in the business."[85] The deal included a large tax write-off of debt for General Dynamics.

Initially, many General Dynamics employees believed that Centaur as an upper stage for Titan IV was the quarry that prompted the Martin Marietta purchase. After several unsuccessful missions with the Titan II-Transtage, the company had withdrawn from the commercial market.[86] However, the company soon recognized that the new Atlas family of vehicles had greater commercial potential than the powerful, but far less technically advanced, Titan. Martin Marietta would later decide to halt production of Titan II and III and declare them obsolete.[87]

The sale of the General Dynamics Space Systems Division to Martin Marietta Astronautics included all of the Atlas-Centaur and Centaur for the Titan IV intellectual property, tooling, inventory, launch-site assets, and work in progress. Martin Marietta swiftly relocated the Space Systems Division to its Waterton Plant south of Denver in the winter of 1995. Manufacture of adapters and fairings remained at the General Dynamics plant in Harlingen, Texas. Tank manufacture, because of the advanced technical skills required for welding, remained at Air Force

---

[83] Mike Patzer, Robert C. White, and Terry A. Bohlen, "Status and Review of the Commercial Atlas Program," 15th AIAA International Communications Satellite Systems Conference, San Diego, February 1994.

[84] Interview with Richard Hauck by Mark Bowles, 23 August 2001.

[85] "Marietta Warns on Acquisition," *New York Times* (19 February 1994); "Martin Marietta To Buy General Dynamics," *Washington Post* (23 December 1993).

[86] Comments on manuscript by Ed Bock, 11 March 2002.

[87] The information in this paragraph is from comments on the manuscript by Ed Bock, 11 March 2002.

Plant 19 in San Diego. Martin Marietta assumed the General Dynamics lease. The entire Kearny Mesa plant in San Diego, including all manufacturing, test, and engineering facilities, was razed.

Most of the engineers supporting these two programs were given job offers, and about half of them chose to make the move. Approximately eight hundred former General Dynamics employees transferred to Denver in 1994. Surprisingly, Martin Marietta offered them most of the top jobs. Mike Wynne was named executive vice-president of all launch vehicles, with the exception of Titan, and Edward Bock became the vice-president of the Atlas program. Edward Squires became the vice-president of all Martin Marietta Astronautics operations, and Robert DiNal became vice-president/division chief engineer. Only Tom Knapp, who became director of the Atlas launch vehicle, came from the ranks of the Martin Marietta Titan program. Knapp played a key role in providing a smooth transition and helping former Atlas program people deal with the Titan culture in Denver. Gene Fourney, a longtime member of NASA's resident office in San Diego, worried that precious engineering expertise would be lost in the move. "You know you had this terrific organization back there, and it was really putting a lot of successful flights together. And then you'd move them, uproot them, and wonder what you're going to do to that record."[88] He was soon reassured.

Martin Marietta did not try to impose control over the program but gave the Atlas managers authority to hire additional people from the Denver area to fill vacancies. In May 1995, production began at the new Atlas and Centaur Final Assembly Building at the Waterton facility in Colorado. In September, the company's International Launch Services opened a new headquarters in San Diego, California. That same month, the last Atlas booster to be built in San Diego came off the assembly line, commemorated by a mass signing of the aft end by former General Dynamics workers.

The takeover of General Dynamics by Martin Marietta, a long-time competitor in the launch vehicle business, proved more positive than many former General Dynamics and Lewis Research Center employees could have imagined. Air Force oversight of Titan had produced a distinctive Titan culture at Martin Marietta. Because Air Force personnel changed frequently, they had less time to become involved in the nuts and bolts of Titan-Centaur. In contrast, NASA had extensive in-house technical expertise and had always worked closely with General Dynamics to solve engineering problems. Although commercialization would break this close tie between government and industry, General Dynamics reaped the many benefits that this unique association had produced over the course of more than thirty years.

The forbearance of Martin Marietta management in allowing former General Dynamics employees to play an important role in the transition paid off for the company. Because the

[88] Interview with Gene Fourney by Virginia Dawson, 21 March 2001.

Called Centaur, the upper stage for the Atlas II is raised to a vertical position on Pad 36A for a launch of a GOES satellite for the National Weather Service system, March 2000. (KSC_00PP_0423)

first launches were successful, the company did not try to impose Titan processes and procedures on Atlas. Mike Benik, at that time head of the NASA resident office, observed, "What
happened was the Atlas culture survived even though they were now in Denver at Martin
Marietta, and in effect, [NASA] ended up dealing with two different companies, all under
one roof."[89] Ed Bock commented, "I think if we'd had a failure shortly after we moved here,
we would have been Titanized at the drop of a hat."[90]

Used to Air Force oversight, the Martin Marietta General Manager was at first visibly
uncomfortable when he was informed that it was not the customer's responsibility to make
the decision to launch. Roy Roberts said, "We were making a decision on our own to push
the button, and all we had to know was that the spacecraft was ready."[91] By the end of 1995,
Atlas-Centaur had accomplished a record-breaking eleven successful launches. Far from
being discontinued, Atlas-Centaur became the division's "cash cow."[92]

A group of young engineers at General Dynamics, mentored by the small cadre of Atlas-
Centaur veterans, continued the tradition of innovation at General Dynamics. They
designed the series of evolved commercial vehicles and persevered after the Atlas I engine
failures and the move to Denver with Martin Marietta. Since the three engine failures, there
has been a 100-percent success record in all fifty-nine flights completed through February
2002, despite significant design modifications in each of the II, IIA, IIAS, IIIA, and IIIB
versions of Atlas.

In 1995, further consolidation of the aerospace industry occurred with the merger between
Martin Marietta and Lockheed. This set the stage for establishing the Lockheed Martin
International Launch Services (ILS), a joint venture owned by Lockheed Martin Corporation
Commercial Launch Services and Lockheed Khrunichev Energia International, with headquar-
ters near Washington, District of Columbia. ILS markets both the Atlas family of vehicles and
the Russian Proton, built by Khrunichev State Research and Production Space Center in
Moscow.

This joint venture in marketing proved less difficult than the next step in collaboration
with the Russians. Atlas III, completed in Denver in 1998, was the culmination of General
Dynamics's ambitious ten-year development program. The company chose the Russian-
built RD-180 engine to power the Atlas III booster because of its superiority in terms of
performance, technology, and cost to its nearest competitor. General Dynamics was so
enthusiastic over its choice of the Russian engine that it was willing to put considerable time

---

[89] Interview with Mike Benik by Virginia Dawson, 9 November 1999.

[90] Interview with Ed Bock by Virginia Dawson, 22 March 2001.

[91] Interview with Roy Roberts by Virginia Dawson, 21 March 2001.

[92] Interview with Matthew Smith by Virginia Dawson, 21 March 2001.

and energy into overcoming various objections from the U.S. State Department. (Atlas III had a performance capability of carrying 9,920 pounds to geosynchronous transfer orbit, or an increase of 1,720 pounds over the Atlas IIAS). It was so powerful that the Atlas booster could again be stretched. The engine was throttleable, a feature that enabled it to provide a range of efficient performance for different payloads.

The Russian booster engines paved the way for utilizing the single-engine Centaur for many missions. Pratt & Whitney upgraded the RL10 engine in many steps to provide 20,800 pounds of thrust for both the dual- and single-engine versions of Atlas. Years earlier, Lewis had laid the foundation for this upgrade when Centaur had demonstrated a capability for long coasts and three, four, or five burns. Particularly significant were tests of Centaur's multiple-restart capability in a zero-gravity environment in NASA's Space Power Facility at Plum Brook prior to the Helios (TC-5) mission in 1976.[93]

The single-engine Centaur increased the inherent reliability of the upper stage dramatically because rocket failures are often related to the engine. Half of the propulsion hardware, including the thrust vector control actuators, propellant feed ducting, pneumatic supply, purge tubing, and controls, could be eliminated. The electromechanical engine control system contributed to a leap in performance. Previously, Centaur engines had been chilled down during the coast phase, after separation from Atlas. Now the Centaur single engine could be cooled down during the Atlas booster phase. This saved precious seconds and conserved fuel, which in turn allowed the vehicle to reach orbit more efficiently.

To achieve a greater propellant load, the length of Centaur's tank was stretched to fit on either an Atlas III or a redesigned 12-foot-diameter Atlas V first stage. Carrying the satellite Echostar, the new stretched Centaur was successfully flown on the first Atlas IIIB flight on 21 February 2002.

Atlas V, the culmination of the development program that began with commercialization, is also the company's response to the Air Force demand in 1996 for a more efficient, less expensive launch vehicle—an Evolved Expendable Launch Vehicle, or EELV. Atlas V reflects the global marketplace in which it operates. The first stage has Russian RD-180 engines. It has a thick, structurally stabilized aluminum tank, reinforced with rings and stringer for the first time. Its interstage adapter was built by Construcciones Aeronauticas S.A. (CASA) of Spain and its composite fairings by Contraves of Switzerland. Its standardized features make it extremely versatile. Depending on the weight of the payload and power requirements, up to five large solid rocket motors can be attached to its common core or three boosters can be strapped together. The boosters are identical to facilitate ease of production.

---

[93] Interview with Bill Johns by Virginia Dawson, 22 March 2001. It should be noted that a 35,000-pound-thrust RL10 derivative was developed for the new hydrogen/oxygen upper stage manufactured by McDonnell Douglas/Boeing for Delta III.

Atlas-Centaur, now simply called Atlas, remains one of the most efficient rocket systems ever built. Systems engineering and technical management processes honed over three decades of NASA-industry collaboration are still used in the commercial Atlas program, although the role of NASA in the new world of commercialization is now greatly diminished. If Atlas-Centaur is still flying, both the space systems division at General Dynamics and its counterpart at NASA Glenn have vanished. Bought by Martin Marietta and then merged with Lockheed, all that remains of the famed missile maker's Kearney Mesa complex in San Diego is a lone administration building. To complete the process of commercialization in 1998, NASA phased out the launch vehicles division in Cleveland. A few of the men and women in the division at Glenn transferred to Kennedy Space Center; others retired or moved into other areas of research and technology. Nevertheless, the unique rocket test facilities conceived and built by NASA Lewis engineers continue to provide the nation with a unique capability.

Taming liquid hydrogen is a striking example of how government research contributed to the development of a new technology. Fifty years ago, when engineers at the Lewis Flight Propulsion Laboratory began experimenting with exotic rocket fuels, they could never have imagined that their work would make a significant contribution to a race with the Soviet Union to land human beings on the Moon. Chance also played a role in the development of liquid-hydrogen technology. Without the aborted development by the Air Force of a high-altitude spy plane in the 1950s, the country would have lacked the technology infrastructure to produce liquid hydrogen in the enormous quantities needed for the space program.

Research and testing of liquid-hydrogen rockets at Lewis Research Center directly influenced the 1959 decision to use liquid hydrogen in the upper stages of the Saturn V. Once the commitment to liquid hydrogen had been made, the development of Centaur became of central importance to proving its feasibility. Because of their expertise in high-energy liquid fuels, engineers at Lewis Research Center served as a clearinghouse between 1959 and 1962 for technical know-how vitally important to the design of liquid-hydrogen-fueled rockets. General Dynamics, Pratt & Whitney, Rocketdyne, and Douglas all benefited from their interaction with Lewis rocket researchers, but whether a liquid-hydrogen upper stage could actually perform with the flawless precision required of spacecraft remained unclear. This was still in doubt as late as October 1962, when management of Centaur was transferred to from Marshall to Lewis.

With the benefit of hindsight, Marshall scientist Ernst Stuhlinger has called liquid hydrogen "one of the most momentous innovations in the history of rockets during the second half of the twentieth century." He credits Abe Silverstein with the courageous decision to use it in Saturn's upper stages. He writes:

> Hydrogen rocket technology played an absolutely decisive role not only in the Saturn-Apollo Moon Project, but also in the Shuttle Project that came to life during the 1970s.

A new, high performance engine, the Space Shuttle Main Engine (SSME), was developed in a joint effort by the G. C. Marshall Space Flight Center and the Rocketdyne Division of North American Aviation. By that time, hydrogen technology for rocket engines had reached its full maturity.[94]

The Centaur saga illustrates how success or failure of a technology ultimately rests on people—the decisions they make and their determination to influence the outcome. To adopt a new technology involved risk. That Centaur ultimately won the backing of President Kennedy himself demonstrates the national importance of this technology. A determined band of government engineers at Lewis persevered in the face of rancorous congressional criticism, bad press, and active opposition by Marshall Space Flight Center. A collaborative relationship built on mutual respect between NASA civil servants and government contractors kept the program alive.

However, if liquid-hydrogen technology moved into NASA's mainstream, neither Centaur nor Lewis Research Center ever did. By the 1970s, Centaur was regarded as yesterday's technology, a dependable workhorse for launching planetary probes and commercial satellites while NASA pushed forward with the development of a reusable Space Shuttle. When the Shuttle's technical limitations for launching planetary missions became clear, NASA turned once again to Centaur. Ironically, while Marshall Space Flight Center actively campaigned for the termination of the Centaur program in the 1960s, it fought unsuccessfully for management of the Shuttle/Centaur program.

Nowhere is the theme of NASA's changing tolerance for risk more forcefully demonstrated than in the tortuous history of Shuttle/Centaur. Lewis engineers had confidence in the systems they designed to protect the astronauts should a mission be aborted, but they had great difficulty in negotiating the politics of the manned spaceflight establishment. After the loss of *Challenger* in 1986, it was clear that the program was doomed.

In the 1970s, when NASA decided to phase out its expendable launch vehicles, the successful commercialization of Centaur was inconceivable. Yet global competition for commercial satellite business and the loss of *Challenger* precipitated a change in national priorities. Today, Atlas-Centaur remains one of the most efficient rocket systems ever built. Although its identity is now submerged within the Atlas, Centaur has proven itself as an upper stage beyond the wildest dreams of one of the great visionaries of the space age. Someday a probe powered by Centaur may yet reach Alpha Centauri, the star that tantalized the rocket's gifted designer Krafft Ehricke.

---

[94] Ernst Stuhlinger, "Enabling Technology for Space Transportation," *The Century of Space Science*, vol. 1 (Dordrecht: Kluwer Academic Publishers, 2001), 73–74.

List of Interviews

Eighty-nine members of the Lewis Research Center Centaur team celebrate the fiftieth Atlas-Centaur Flight, June 1979: Joe Kubancik, Gil Widra, Jack Herman, Steve Szabo, Norm Weisberg, Bill Groesbeck, Maurie Duddout, Dick Woelfle, Henry Synor, Lillian Manning, Ed Ziemba, Larry Ross, Jim Patterson, Maynard Weston, Dick Heath, Tom Hill, Joe Nieberding, Sharon Huber, Bob Edwards, Floyd Smith, Barbara Troyan, Konnie Semenchuk, Carl Wentworth, Ted Gerus, Jim Magrini, Don Garman, Paul Kuebeler, Dean Bitler, Bill Kress, Sandy Tingley, Carl Monnin, George Schaefer, Harlan Simon, Peggy Schuler, Rita Hart, Irene Blanchard, Steve Kiacz, Bob Firestone, Bill Misichko, Bill Hulzman, Harry Groth, Tom Seeholzer, Bob Reinberger, Jim Stoll, Baxter Beaton, Merle Jones, Dick Kalo, John Kramer, Bill Middendorf, Gene Cieslewicz, Ed Procasky, Jack Farley, Pam Foerster, John Feagan, Sandy Hanes, Pat Miller, Jack Estes, John Gibb, Ed Muckley, Bob Miller, Dick Flage, Dick Orzechowski, John Roberts, Charley Eastwood, Ralph Kuivinen, Frank Gue, Janos Borsody, Ken Baud, Gene Fourney, Ray Salmi, John Nechvatal, Bob Jabo, Ed Jeris, Al Hahn, Cecil O'Dear, Andy Dobos, Gus Delaney, Bob Lubick, Steve Szpatura, Ted Porada, Art Zimmerman, John Riehl, Ray Lacovic, Bob Rohal, John Bulloch, Jim McAleese, Chuck Tiede, John Catone, and Earl Hanes. (Glenn Research Center unprocessed photo)

Glenn Research Center
    Brun, Rinaldo J., by Virginia Dawson, 17 June 1999
    Everett, Ronald, by V. Dawson, 17 December 2001
    Fourney, Eugene, by V. Dawson, 21 March 2001 (at Lockheed Martin, Denver)
    Geye, Richard, by V. Dawson, 27 May 1999

Himmel, Seymour, by V. Dawson, 1 March 2000

Kozar, Robert, by V. Dawson and Mark Bowles (at Plum Brook)

Lundin, Bruce, by V. Dawson, 7 March 2000

Muckley, Edwin, by V. Dawson and M. Bowles, 29 April 1999

Nettles, Cary, by V. Dawson, 4 March 1999

Nieberding, Joe, by V. Dawson and M. Bowles, 15 April 1999, 9 May 2000, 7 November 2001

Reshotko, Meyer, by V. Dawson, 17 December 2001 (Cleveland Heights)

Robbins, Red, by M. Bowles, 21 March 2000

Ross, Larry, by M. Bowles, 1 March 2000

Simon, Harlan, by V. Dawson, 20 March 1985

Spurlock, Frank, by V. Dawson, 6 April 1999

Stofan, Andrew, by M. Bowles, 13 April 2000 (by phone)

Swavely, Jim, by V. Dawson, 16 November 1999

Weyers, Vern, by M. Bowles, 8 April 2000

Ziemba, Edmund R., by V. Dawson, 19 May 1999

Zimmerman, Art, by V. Dawson and M. Bowles, 5 August 1999

Johnson Space Center

Hauck, Rick, by M. Bowles, 23 August 2001 (by phone)

Kohrs, Richard, by M. Bowles, 13 July 2001 (by phone)

Kennedy Space Center

Benik, Mike, by V. Dawson, 9 November 1999

Carpenter, Mary Sue, by V. Dawson, 11 November 1999

Curington, Floyd, by V. Dawson, 9 November 1999

François, Steve, by V. Dawson, 9 November 1999

Gossett, John, by V. Dawson, 2 July 2002

Gray, Robert, by V. Dawson, 9 November 1999

Mackey, Arthur (Skip), by V. Dawson, 9 November 1999

Neilon, John, by V. Dawson, 1 July 2002

Womack, James, by V. Dawson, 11 November 1999

Former General Dynamics employees, San Diego

Anthony, Frank, by V. Dawson, 6 June 2000

Benzwi, Robert, by V. Dawson, 7 June 2000

Bonesteel, Howard, by V. Dawson, 7 June 2000

Bradley, Robert, by V. Dawson, 7 June 2000

Chamberlain, Stanley, by V. Dawson, 7 June 2000

Chandik, Thomas, by V. Dawson, 6 June 2000

Hansen, Grant, by V. Dawson, 6 June 2000

Kachigan, Karl, by V. Dawson, 7 June 2000

Kaukonen, Everett, by V. Dawson, 6 June 2000

Laird, Hugh, by V. Dawson, 6 June 2000

Lesney, Donald, by V. Dawson, 5 June 2000

Lovelace, Alan, by V. Dawson, 19 July 2002 (by phone)

Lynch, Roger, by V. Dawson, 5 June 2000

Martin, Richard, by V. Dawson, 5 June 2000

Merino, Frederick, by V. Dawson, 5 June 2000
Perry, L. Charles, by V. Dawson, 6 June 2000
Wilmot, Allan, by V. Dawson, 6 June 2000
Wilson, Charles (Chuck), by 7 June 2000
Winkler, Martin, by V. Dawson, 21 March 2001
Zeenkov, Seymour, by V. Dawson, 5 June 2000

Lockheed Martin, Denver, Colorado
Bock, Edward, by V. Dawson, 22 March 2001
Jensen, Michael, by V. Dawson, 22 March 2001
Johns, William, by V. Dawson, 22 March 2001
Owara, Vern, by V. Dawson, 22 March 2001
Reynolds, Robert, by V. Dawson, 21 March 2001
Roberts, Roy, by V. Dawson, 21 March 2001
Smith, Matthew, by V. Dawson, 21 March 2001

JPL
Haynes, Norm, by V. Dawson, 9 June 2000
O'Neil, William, by V. Dawson, 9 June 2000
Barnett, Phillip, by V. Dawson, 9 June 2000
Kohlhase, Charles, by V. Dawson, 9 June 2000
Shaw, Lutha (Tom), by V. Dawson, 10 November 1999 (at Kennedy Space Center)

About the Authors

Virginia P. Dawson is the author of *Nature's Enigma, Engines and Innovation: Lewis Laboratory and American Propulsion Technology,* and *Lincoln Electric: a History* and has contributed chapters in several collections on aerospace history. She is founder and president of History Enterprises, Inc., *http://www.HistoryEnterprises.com*, and an adjunct professor of history at Case Western Reserve University in Cleveland, Ohio.

Mark D. Bowles received his B.A. in psychology and M.A. in history from the University of Akron. He earned his Ph.D. from Case Western Reserve University in history in 1999. He has also been a Tomash Fellow from the Charles Babbage Institute at the University of Minnesota. Dr. Bowles is the author of *Our Healing Mission*, a history of Saint Francis Hospital and Medical Center in Hartford, Connecticut. He is currently writing *Reactor in the Garden*, a history of NASA's nuclear research reactor at Plum Brook Station.

The NASA History Series

Reference Works, NASA SP-4000

Grimwood, James M. *Project Mercury: A Chronology*. NASA SP-4001, 1963.

Grimwood, James M., and C. Barton Hacker, with Peter J. Vorzimmer. *Project Gemini Technology and Operations: A Chronology*. NASA SP-4002, 1969.

Link, Mae Mills. *Space Medicine in Project Mercury*. NASA SP-4003, 1965.

Astronautics and Aeronautics, 1963: Chronology of Science, Technology, and Policy. NASA SP-4004, 1964.

Astronautics and Aeronautics, 1964: Chronology of Science, Technology, and Policy. NASA SP-4005, 1965.

Astronautics and Aeronautics, 1965: Chronology of Science, Technology, and Policy. NASA SP-4006, 1966.

Astronautics and Aeronautics, 1966: Chronology of Science, Technology, and Policy. NASA SP-4007, 1967.

Astronautics and Aeronautics, 1967: Chronology of Science, Technology, and Policy. NASA SP-4008, 1968.

Ertel, Ivan D., and Mary Louise Morse. *The Apollo Spacecraft: A Chronology, Volume I, Through November 7, 1962*. NASA SP-4009, 1969.

Morse, Mary Louise, and Jean Kernahan Bays. *The Apollo Spacecraft: A Chronology, Volume II, November 8, 1962–September 30, 1964*. NASA SP-4009, 1973.

Brooks, Courtney G., and Ivan D. Ertel. *The Apollo Spacecraft: A Chronology, Volume III, October 1, 1964-January 20, 1966*. NASA SP-4009, 1973.

Ertel, Ivan D., and Roland W. Newkirk, with Courtney G. Brooks. *The Apollo Spacecraft: A Chronology, Volume IV, January 21, 1966–July 13, 1974*. NASA SP-4009, 1978.

Astronautics and Aeronautics, 1968: Chronology of Science, Technology, and Policy. NASA SP-4010, 1969.

Newkirk, Roland W., and Ivan D. Ertel, with Courtney G. Brooks. *Skylab: A Chronology*. NASA SP-4011, 1977.

Van Nimmen, Jane, and Leonard C. Bruno, with Robert L. Rosholt. *NASA Historical Data Book, Volume I: NASA Resources, 1958–1968*. NASA SP-4012, 1976, rep. ed. 1988.

Ezell, Linda Neuman. *NASA Historical Data Book, Volume II: Programs and Projects, 1958–1968*. NASA SP-4012, 1988.

Ezell, Linda Neuman. *NASA Historical Data Book, Volume III: Programs and Projects, 1969–1978*. NASA SP-4012, 1988.

Gawdiak, Ihor Y., with Helen Fedor, compilers. *NASA Historical Data Book, Volume IV: NASA Resources, 1969–1978*. NASA SP-4012, 1994.

Rumerman, Judy A., compiler. *NASA Historical Data Book, 1979-1988: Volume V, NASA Launch Systems, Space Transportation, Human Spaceflight, and Space Science*. NASA SP-4012, 1999.

Rumerman, Judy A., compiler. *NASA Historical Data Book, Volume VI: NASA Space Applications, Aeronautics and Space Research and Technology, Tracking and Data Acquisition/Space Operations, Commercial Programs, and Resources, 1979–1988*. NASA SP-2000-4012, 2000.

Astronautics and Aeronautics, 1969: Chronology of Science, Technology, and Policy. NASA SP-4014, 1970.

Astronautics and Aeronautics, 1970: Chronology of Science, Technology, and Policy. NASA SP-4015, 1972.

Astronautics and Aeronautics, 1971: Chronology of Science, Technology, and Policy. NASA SP-4016, 1972.

Astronautics and Aeronautics, 1972: Chronology of Science, Technology, and Policy. NASA SP-4017, 1974.

Astronautics and Aeronautics, 1973: Chronology of Science, Technology, and Policy. NASA SP-4018, 1975.

Astronautics and Aeronautics, 1974: Chronology of Science, Technology, and Policy. NASA SP-4019, 1977.

Astronautics and Aeronautics, 1975: Chronology of Science, Technology, and Policy. NASA SP-4020, 1979.

Astronautics and Aeronautics, 1976: Chronology of Science, Technology, and Policy. NASA SP-4021, 1984.

Astronautics and Aeronautics, 1977: Chronology of Science, Technology, and Policy. NASA SP-4022, 1986.

Astronautics and Aeronautics, 1978: Chronology of Science, Technology, and Policy. NASA SP-4023, 1986.

Astronautics and Aeronautics, 1979–1984: Chronology of Science, Technology, and Policy. NASA SP-4024, 1988.

Astronautics and Aeronautics, 1985: Chronology of Science, Technology, and Policy. NASA SP-4025, 1990.

Noordung, Hermann. *The Problem of Space Travel: The Rocket Motor.* Edited by Ernst Stuhlinger and J. D. Hunley, with Jennifer Garland. NASA SP-4026, 1995.

Astronautics and Aeronautics, 1986–1990: A Chronology. NASA SP-4027, 1997.

Astronautics and Aeronautics, 1990–1995: A Chronology. NASA SP-2000-4028, 2000.

Management Histories, NASA SP-4100

Rosholt, Robert L. *An Administrative History of NASA, 1958–1963.* NASA SP-4101, 1966.

Levine, Arnold S. *Managing NASA in the Apollo Era.* NASA SP-4102, 1982.

Roland, Alex. *Model Research: The National Advisory Committee for Aeronautics, 1915–1958.* NASA SP-4103, 1985.

Fries, Sylvia D. *NASA Engineers and the Age of Apollo.* NASA SP-4104, 1992.

Glennan, T. Keith. *The Birth of NASA: The Diary of T. Keith Glennan.* J. D. Hunley, ed. NASA SP-4105, 1993.

Seamans, Robert C., Jr. *Aiming at Targets: The Autobiography of Robert C. Seamans, Jr.* NASA SP-4106, 1996.

Garber, Stephen J., *Looking Backward, Looking Forward: Forty Years of U.S. Human Spaceflight Symposium.* NASA SP-4107, 2002.

Project Histories, NASA SP-4200

Swenson, Loyd S., Jr., James M. Grimwood, and Charles C. Alexander. *This New Ocean: A History of Project Mercury.* NASA SP-4201, 1966; rep. ed. 1998.

Green, Constance McLaughlin, and Milton Lomask. *Vanguard: A History.* NASA SP-4202, 1970; rep. ed. Smithsonian Institution Press, 1971.

Hacker, Barton C., and James M. Grimwood. *On Shoulders of Titans: A History of Project Gemini.* NASA SP-4203, 1977.

Benson, Charles D., and William Barnaby Faherty. *Moonport: A History of Apollo Launch Facilities and Operations*. NASA SP-4204, 1978.

Brooks, Courtney G., James M. Grimwood, and Loyd S. Swenson, Jr. *Chariots for Apollo: A History of Manned Lunar Spacecraft*. NASA SP-4205, 1979.

Bilstein, Roger E. *Stages to Saturn: A Technological History of the Apollo/Saturn Launch Vehicles*. NASA SP-4206, 1980, rep. ed. 1997.

SP-4207 not published.

Compton, W. David, and Charles D. Benson. *Living and Working in Space: A History of Skylab*. NASA SP-4208, 1983.

Ezell, Edward Clinton, and Linda Neuman Ezell. *The Partnership: A History of the Apollo-Soyuz Test Project*. NASA SP-4209, 1978.

Hall, R. Cargill. *Lunar Impact: A History of Project Ranger*. NASA SP-4210, 1977.

Newell, Homer E. *Beyond the Atmosphere: Early Years of Space Science*. NASA SP-4211, 1980.

Ezell, Edward Clinton, and Linda Neuman Ezell. *On Mars: Exploration of the Red Planet, 1958–1978*. NASA SP-4212, 1984.

Pitts, John A. *The Human Factor: Biomedicine in the Manned Space Program to 1980*. NASA SP-4213, 1985.

Compton, W. David. *Where No Man Has Gone Before: A History of Apollo Lunar Exploration Missions*. NASA SP-4214, 1989.

Naugle, John E. *First Among Equals: The Selection of NASA Space Science Experiments*. NASA SP-4215, 1991.

Wallace, Lane E. *Airborne Trailblazer: Two Decades with NASA Langley's Boeing 737 Flying Laboratory*. NASA SP-4216, 1994.

Butrica, Andrew J., ed. *Beyond the Ionosphere: Fifty Years of Satellite Communication*. NASA SP-4217, 1997.

Butrica, Andrew J. *To See the Unseen: A History of Planetary Radar Astronomy*. NASA SP-4218, 1996.

Mack, Pamela E., ed. *From Engineering Science to Big Science: The NACA and NASA Collier Trophy Research Project Winners*. NASA SP-4219, 1998.

Reed, R. Dale, with Darlene Lister. *Wingless Flight: The Lifting Body Story*. NASA SP-4220, 1997.

Heppenheimer, T. A. *The Space Shuttle Decision: NASA's Search for a Reusable Space Vehicle*. NASA SP-4221, 1999.

Hunley, J. D., ed. *Toward Mach 2: The Douglas D-558 Program*. NASA SP-4222, 1999.

Swanson, Glen E., ed. *"Before this Decade is Out . . .": Personal Reflections on the Apollo Program*. NASA SP-4223, 1999.

Tomayko, James E. *Computers Take Flight: A History of NASA's Pioneering Digital Fly-by-Wire Project*. NASA SP-2000-4224, 2000.

Morgan, Clay. *Shuttle-Mir: The U.S. and Russia Share History's Highest Stage*. NASA SP-2001-4225, 2001.

Mudgway, Douglas J. *Uplink-Downlink: A History of the Deep Space Network, 1957–1997*. NASA SP-2001-4227, 2002.

Center Histories, NASA SP-4300

Rosenthal, Alfred. *Venture into Space: Early Years of Goddard Space Flight Center.* NASA SP-4301, 1985.

Hartman, Edwin P. *Adventures in Research: A History of Ames Research Center, 1940–1965.* NASA SP-4302, 1970.

Hallion, Richard P. *On the Frontier: Flight Research at Dryden, 1946-1981.* NASA SP-4303, 1984.

Muenger, Elizabeth A. *Searching the Horizon: A History of Ames Research Center, 1940–1976.* NASA SP-4304, 1985.

Hansen, James R. *Engineer in Charge: A History of the Langley Aeronautical Laboratory, 1917–1958.* NASA SP-4305, 1987.

Dawson, Virginia P. *Engines and Innovation: Lewis Laboratory and American Propulsion Technology.* NASA SP-4306, 1991.

Dethloff, Henry C. *"Suddenly Tomorrow Came . . .": A History of the Johnson Space Center.* NASA SP-4307, 1993.

Hansen, James R. *Spaceflight Revolution: NASA Langley Research Center from Sputnik to Apollo.* NASA SP-4308, 1995.

Wallace, Lane E. *Flights of Discovery: 50 Years at the NASA Dryden Flight Research Center.* NASA SP-4309, 1996.

Herring, Mack R. *Way Station to Space: A History of the John C. Stennis Space Center.* NASA SP-4310, 1997.

Wallace, Harold D., Jr. *Wallops Station and the Creation of the American Space Program.* NASA SP-4311, 1997.

Wallace, Lane E. *Dreams, Hopes, Realities: NASA's Goddard Space Flight Center, The First Forty Years.* NASA SP-4312, 1999.

Dunar, Andrew J., and Stephen P. Waring. *Power to Explore: A History of the Marshall Space Flight Center.* NASA SP-4313, 1999.

Bugos, Glenn E. *Atmosphere of Freedom: Sixty Years at the NASA Ames Research Center.* NASA SP-2000-4314, 2000.

General Histories, NASA SP-4400

Corliss, William R. *NASA Sounding Rockets, 1958–1968: A Historical Summary.* NASA SP-4401, 1971.

Wells, Helen T., Susan H. Whiteley, and Carrie Karegeannes. *Origins of NASA Names.* NASA SP-4402, 1976.

Anderson, Frank W., Jr. *Orders of Magnitude: A History of NACA and NASA, 1915–1980.* NASA SP-4403, 1981.

Sloop, John L. *Liquid Hydrogen as a Propulsion Fuel, 1945–1959.* NASA SP-4404, 1978.

Roland, Alex. *A Spacefaring People: Perspectives on Early Spaceflight.* NASA SP-4405, 1985.

Bilstein, Roger E. *Orders of Magnitude: A History of the NACA and NASA, 1915–1990.* NASA SP-4406, 1989.

Logsdon, John M., ed., with Linda J. Lear, Jannelle Warren-Findley, Ray A. Williamson, and Dwayne A. Day. *Exploring the Unknown: Selected Documents in the History of the U.S. Civil Space Program, Volume I, Organizing for Exploration.* NASA SP-4407, 1995.

Logsdon, John M., ed., with Dwayne A. Day and Roger D. Launius. *Exploring the Unknown: Selected Documents in the History of the U.S. Civil Space Program, Volume II, Relations with Other Organizations.* NASA SP-4407, 1996.

Logsdon, John M., ed., with Roger D. Launius, David H. Onkst, and Stephen J. Garber. *Exploring the Unknown: Selected Documents in the History of the U.S. Civil Space Program, Volume III, Using Space.* NASA SP-4407, 1998.

Logsdon, John M., gen. ed., with Ray A. Williamson, Roger D. Launius, Russell J. Acker, Stephen J. Garber, and Jonathan L. Friedman. *Exploring the Unknown: Selected Documents in the History of the U.S. Civil Space Program, Volume IV, Accessing Space.* NASA SP-4407, 1999.

Logsdon, John M., gen. ed., with Amy Paige Snyder, Roger D. Launius, Stephen J. Garber, and Regan Anne Newport. *Exploring the Unknown: Selected Documents in the History of the U.S. Civil Space Program, Volume V, Exploring the Cosmos.* NASA SP 2001 4407, 2001.

Siddiqi, Asif A. *Challenge to Apollo: The Soviet Union and the Space Race, 1945–1974.* NASA SP-2000-4408, 2000.

Monographs in Aerospace History, NASA SP-4500

Launius, Roger D., and Aaron K. Gillette, comps. *Toward a History of the Space Shuttle: An Annotated Bibliography.* Monograph in Aerospace History, No. 1, 1992. Out of print.

Launius, Roger D., and J. D. Hunley, comps. *An Annotated Bibliography of the Apollo Program.* Monograph in Aerospace History, No. 2, 1994.

Launius, Roger D. *Apollo: A Retrospective Analysis.* Monograph in Aerospace History, No. 3, 1994.

Hansen, James R. *Enchanted Rendezvous: John C. Houbolt and the Genesis of the Lunar-Orbit Rendezvous Concept.* Monograph in Aerospace History, No. 4, 1995.

Gorn, Michael H. *Hugh L. Dryden's Career in Aviation and Space.* Monograph in Aerospace History, No. 5, 1996.

Powers, Sheryll Goecke. *Women in Flight Research at NASA Dryden Flight Research Center from 1946 to 1995.* Monograph in Aerospace History, No. 6, 1997.

Portree, David S. F., and Robert C. Trevino. *Walking to Olympus: An EVA Chronology.* Monograph in Aerospace History, No. 7, 1997.

Logsdon, John M., moderator. *Legislative Origins of the National Aeronautics and Space Act of 1958: Proceedings of an Oral History Workshop.* Monograph in Aerospace History, No. 8, 1998.

Rumerman, Judy A., comp. *U.S. Human Spaceflight, A Record of Achievement 1961–1998.* Monograph in Aerospace History, No. 9, 1998.

Portree, David S. F. *NASA's Origins and the Dawn of the Space Age.* Monograph in Aerospace History, No. 10, 1998.

Logsdon, John M. *Together in Orbit: The Origins of International Cooperation in the Space Station.* Monograph in Aerospace History, No. 11, 1998.

Phillips, W. Hewitt. *Journey in Aeronautical Research: A Career at NASA Langley Research Center.* Monograph in Aerospace History, No. 12, 1998.

Braslow, Albert L. *A History of Suction-Type Laminar-Flow Control with Emphasis on Flight Research.* Monograph in Aerospace History, No. 13, 1999.

Logsdon, John M., moderator. *Managing the Moon Program: Lessons Learned From Apollo.* Monograph in Aerospace History, No. 14, 1999.

Perminov, V. G. *The Difficult Road to Mars: A Brief History of Mars Exploration in the Soviet Union.* Monograph in Aerospace History, No. 15, 1999.

Tucker, Tom. *Touchdown: The Development of Propulsion Controlled Aircraft at NASA Dryden.* Monograph in Aerospace History, No. 16, 1999.

Maisel, Martin D., Demo J. Giulianetti, and Daniel C. Dugan. *The History of the XV-15 Tilt Rotor Research Aircraft: From Concept to Flight.* NASA SP-2000-4517, 2000.

Jenkins, Dennis R. *Hypersonics Before the Shuttle: A Concise History of the X-15 Research Airplane.* NASA SP-2000-4518, 2000.

Chambers, Joseph R. *Partners in Freedom: Contributions of the Langley Research Center to U.S. Military Aircraft in the 1990s.* NASA SP-2000-4519, 2000.

Waltman, Gene L. *Black Magic and Gremlins: Analog Flight Simulations at NASA's Flight Research Center.* NASA SP-2000-4520, 2000.

Portree, David S. F. *Humans to Mars: Fifty Years of Mission Planning, 1950–2000.* NASA SP-2001-4521, 2001.

Thompson, Milton O., with J. D. Hunley. *Flight Research: Problems Encountered and What They Should Teach Us.* NASA SP-2000-4522, 2000.

Tucker, Tom. *The Eclipse Project.* NASA SP-2000-4523, 2000.

Siddiqi, Asif A. *Deep Space Chronicle: A Chrononology of Deep Space and Planetary Probes 1958–2000.* NASA SP-2002-4524, 2002.

Merlin, Peter W. *Mach 3+: NASA/USAF YF-12 Flight Research, 1969–1979.* NASA SP-2001-4525, 2001.

Anderson, Seth B. *Memoirs of an Aeronautical Engineer: Flight Test at Ames Research Center: 1940–1970.* NASA SP-2002-4526, 2002.

Renstrom, Arthur G. *Wilbur and Orville Wright: A Bibliography Commemorating the One-Hundredth Anniversary of the First Powered Flight on December 17, 1903.* NASA SP-2002-4527, 2002.

There is no monograph 28.

Chambers, Joseph R. *Concept to Reality: Contributions of the NASA Langley Research Center to U.S. Civil Aircraft of the 1990s.* NASA SP-2003-4529, 2003.

Peebles, Curtis, ed. *The Spoken Word: Recollections of Dryden History, The Early Years.* NASA SP-2003-4530, 2003.

Jenkins, Dennis R., Tony Landis, and Jay Miller. *American X-Vehicles: An Inventory–X-1 to X-50.* NASA SP-2003-4531, 2003.

Renstrom, Arthur G. *Wilbur and Orville Wright: A Reissue of a Chronology Commemorating the Hundredth Anniversary of the Birth of Orville Wright, August 19, 1871.* NASA SP-2003-4532, 2003.

INDEX

A

ADDJUST steering program, 73, 128, 130, 144

Advanced Research Projects Agency (ARPA), iii, 13, 18, 20, 22

Advent Communications Satellite Program, 22, 27, 29, 32, 51, 67

Aerobee rocket, 27

Aerojet Engineering Corporation, 14, 15, 142

Aerospace Management, 66

Agena upper stage rocket, x, 22, 25, 26, 51, 52, 54, 55, 56, 51, 93. *See also* Atlas-Agena, Thor-Agena

Air Force, ix, 13, 14, 18, 22, 26, 33, 40, 43, 44, 47, 48, 50, 62, 64, 71, 86, 130, 141, 144, 151, 168, 170, 172, 174, 176, 180, 184, 193, 194, 196, 205, 209, 214, 215; and commercialization, 226, 232, 235, 236, 246

Aldridge, Edward (Pete), 235, 236

Aldrin, Edwin (Buzz), 93

American Rocket Society, 2

Ames Research Center, 62, 107

Analex, Inc., 209, 213

Andrews, Jan, 68

Angara engine, v

Anthony, Frank, 67, 68, 128, 256

Apollo 4, 96

Apollo 11, 100, 101

Apollo program, xi, 33, 50, 56, 64, 82, 83, 88, 89, 94, 100, 101, 163, 168, 174, 205

Applications Technology Satellite (ATS), 108, 110, 111, 113, 116, 123

Appold, Norman, 18

Ariane (launch vehicle), iv, v, vii, ix, 221, 228, 231, 232, 233, 236, 242, 243, 246

Arianespace, 221, 233, 235, 242

Army Ballistic Missile Agency, 3

Arnold Engineering Development Center, 27

Asteroid 29 Amphitrite, 190, 191

Asteroid Gaspra, 211

Astrotech International, 192

AT&T, 223, 225

Atlantis, 202

Atlas (launch vehicle), iv, v, ix, 1, 18, 20, 22, 38–39, 49–51, 60–61, 74, 76–78, 85, 124, 126, 178, 189

Atlas (missile), 1, 10, 39, 7–12, 39

Atlas G, 232, 242

Atlas I, 242, 246

Atlas II, 242, 243, 249, 250

Faraday, Michael, 4
Farley, Jack, 255
Feagan, John, 255
Federal Ministry for Research and Technology, 147
Files, Colonel William, 194
Firestone, Robert, 255
Flage, Richard, 68, 255
Fleet Satellite Communications. *See* FLTSATCOM
Fleming, William, 61, 82
Fletcher, James C., 101, 210
Flippo, Ronnie G., 174, 176
FLTSATCOM (Fleet Satellite Communications), 106, 108
FLTSATCOM satellite, 222
FLTSATCOM F1, 223
FLTSATCOM F2, 223
FLTSATCOM F3, 223
FLTSATCOM F4, 223, 226
FLTSATCOM F5, 223
FLTSATCOM F6, 224, 234
FLTSATCOM F7, 224
FLTSATCOM F8, 224, 238
Foerster, Pam, 255
Ford Aerospace and Communications Corp., 224
Forsythe, D. L., 25
Fourney, Eugene, 248, 255
Foushee, Robert, 68
François, Steve, 256
Frau im Mond, 1
Friedrich, Hans, 4
Frosch, Robert A., 136, 175

G
Gabriel, David, 62–64, 66, 93
Galileo mission, vii, 163, 172, 177, 180, 190–192, 194, 196, 198, 202, 204–205, 207–212
Galilei, Galileo, 154
Gardner, Trevor, 8
Garman, Don, 126, 255
Garside, Joseph, 68
Gemini mission, 141
General Accounting Office (GAO), 51, 61–62

Hunley, John, xii
Huygens, Christian, 154
hydrogen liquifier (cryostat), 14, 17

I
Inertial Upper Stage (IUS), 171–176, 178, 190, 192, 211
Integrate, Transfer and Launch (ITL) facility, 141
integrating Centaur with Atlas, 39–40, 76, 77, 230
integrating Centaur with Shuttle, 178, 180, 185, 195, 198
integrating Centaur with Titan, 143, 144, 163
Intelsat. *See* International Telecommunications Satellite Organization
Intelsat III–IV, 224
Intelsat IV, 223
Intelsat IVA, 223
Intelsat V, 223, 228, 242
Intelsat VA, 224
Intelsat VII, 241
Interim Orbital Transfer Vehicle (IOTV), 171
International Solar Polar Mission. *See* Ulysses spacecraft
International Telecommunications Satellite Organization (Intelsat), 106, 108, 125, 131, 138, 149, 155, 222, 223, 226, 228
Irene Blanchard, 255
Italian Space Agency (Agenzia Spaziale Italiana), 215, 242

J
J-2 Rocketdyne engine, 54, 85, 96, 98
Jabo, Robert, 255
Jakobowski, Walter, 151
James, J. N., 104
Jenkins, Dennis R., xi
Jensen, Michael, 257
Jeris, Ed, 255
Jet Propulsion Laboratory (JPL), 14, 15, 26, 29, 33, 52, 53, 55, 57, 59, 60, 67–69, 83, 91, 92, 95, 104, 105, 107, 120, 157, 159, 165, 190, 191, 194, 197, 211, 212, 226
Johns, William, 257
Johnson Space Center (JSC), vii, viii, 168, 174, 177–180, 182, 184, 185, 191, 194, 196–202, 207–209, 218
Johnson, Clarence (Kelly), 17, 18
Johnson, President Lyndon, 100
Johnson, Roy, 11
Johnson, Vincent, 53, 54, 73, 125, 226
Johnston, Herrick L., 13, 14
Jonash, Edmund, 62, 63, 64, 93

O

O'Dear, Cecil, 255
O'Neil, William, 69, 120, 257
Oberth, Hermann, 1, 5, 8
Office of Commercial Space Transportation, 222
Office of Management and Budget (OMB), 173, 174, 176
Office of Space Flight Programs, 18, 23
Ohio State University, 13, 14
Opp, Al, 152
Orbital Sciences Corporation, 192
Orbiting Astronomical Observatory (OAO), 108, 113, 121, 123, 136, 138
Orbiting Astronomical Observatory-2, 113, 116, 137
Orbiting Astronomical Observatory-3, 113, 116
Orbiting Astronomical Observatory-4, 116, 117
Ordin, Paul, 15
Orzechowski, Richard, 255
Ostrander, Don, 26, 31, 33, 42
Owara, Vern, 257

P

Palley, I. Nelvin, 86
parking orbit. *See* Centaur coast period
Patterson, James, 126, 226, 255
Patterson, William, 4, 12
payload (Shuttle), 198–200
Payload Integration Plan, 193
Peery, David, 73
Penner, Rudolph, 231
Performance Trajectory Group (PTG), 67–68
Perkins, Clay, 84
Perry, L. (Len) Charles, xii, 65, 257
Philadelphia Evening Bulletin, 97
Pickering, William, 82
Pioneer 4, 45
Pioneer 10, 116, 123–125, 128, 130
Pioneer 11, 116, 124, 125, 128, 130, 131
Pioneer Venus, 108, 116, 133, 134, 136
Plain Dealer, 88, 231
planetary assists, 162, 175, 192, 208, 211, 212
Plum Brook Spacecraft Propulsion Research Facility (B-2), 73, 125–127, 143, 146
Plum Brook Station, Sandusky, Ohio, 65, 73, 251, 104, 106, 121, 138, 146
Point Loma test facility, 20, 22, 65, 106

Shuttle/Centaur: advantages, 176; cancellation, 168, 235, 237; fuel dump, 175, 185, 200, 206, 207, 216; mission patch, 169; rollout, 203–205; safety, 170, 173–177, 186, 189, 196, 198, 200–202, 206, 207, 209, 215, 217, 253; weight issues, 208, 209

Shuttle/Centaur Project Office, 182, 191, 194

Shuttle/Centaur-1, 204–206

Shuttle/Centaur-2, 205, 206

Shuttle-Centaur man-rating, 169, 178, 201

Siddiqi, Asif, 98

Silverstein, Abe, vi, ix, 12, 13, 18, 32, 33, 34, 99, 100, 237, 252; and direct ascent, 66–67; and Lewis management of Centaur, 53–54, 57, 59, 62, 65, 71–75, 77, 92–93; and Saturn, 23–25

Simon, Harlan, 59, 64, 65, 255, 256

Slayton, Deke, 93

Slone, Henry, 106

Sloop, John, ix, 5, 6, 15, 34, 35

Smith, Clyde Curry, 122

Smith, Dick G., 178

Smith, Floyd, 214, 255

Smith, Matthew, 257

SOHO (Solar and Heliospheric Observatory), 242

Solid Motor Assembly Building (SMAB), 141, 142

Solid Rocket Motors, 124, 142

Solomon, George, 66

Soviet Union, v, 3, 4, 11, 47, 98, 222; and space race, 104, 105, 134, 168

Space Commerce, 245

Space Plasma High Voltage Interaction Experiment (SPHINX), 145

Space Power Chamber, 71

space science, 103, 147, 152, 156, 162, 163, 165, 191, 192

Space Shuttle limitations, 169

Space Shuttle Main Engine (SSME), 255

Space Shuttle, iv, v, vi, 101, 102, 221, 222, 226, 228, 229, 231, 232, 233, 239, 243, 253

Space Taskforce Group, 169

Space Technology Laboratories (STL), 60, 66

Space Transportation System (STS). *See* Space Shuttle

space tug, 170, 171

Space, 5

Spacecraft Assembly and Encapsulation Facility, 161

Sparks, Brian, 52, 54, 55

Spurlock, Lewis Omer (Frank), xii, 107, 108, 183, 245, 256

Sputnik, 3, 11, 103, 222

Squires, Edward, 248

Stafford, Thomas, 93

Wilmot, Allan, 257
Wilson, Chuck, 257
winds, upper altitude, 124, 128, 130. *See also* ADDJUST steering program
Winkler, Martin, xii, 189, 193, 196, 216, 217, 219, 243, 257
Winslow, Paul, 106
Woelfle, Richard, 255
Womack, James, 44, 238, 256
Wright-Patterson Air Force Base, 17, 18
Wynne, Michael, 247, 248

Y
Yardley, John, 178, 179
York, Herbert, iii, 13, 22
Young, John, 207

Z
Zeenkov, Seymour, 257
Ziemba, Edmund R., 120, 130, 255, 256
Zimmerman, Art, xii, 66, 68, 69, 255, 256
Zucrow, Maurice, 6